Introduction to Solar Photovoltaics

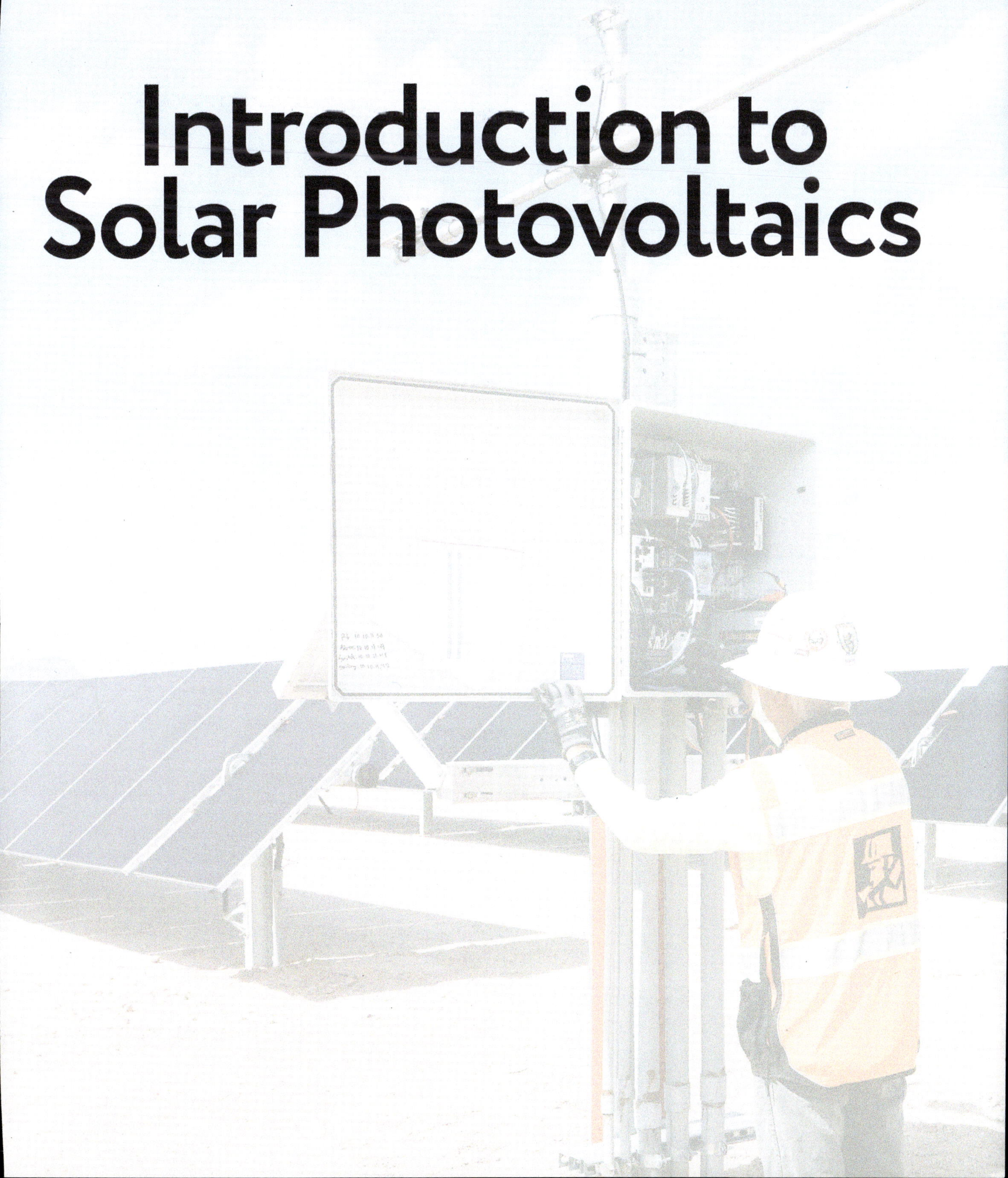

Introduction to Solar Photovoltaics

SECOND EDITION

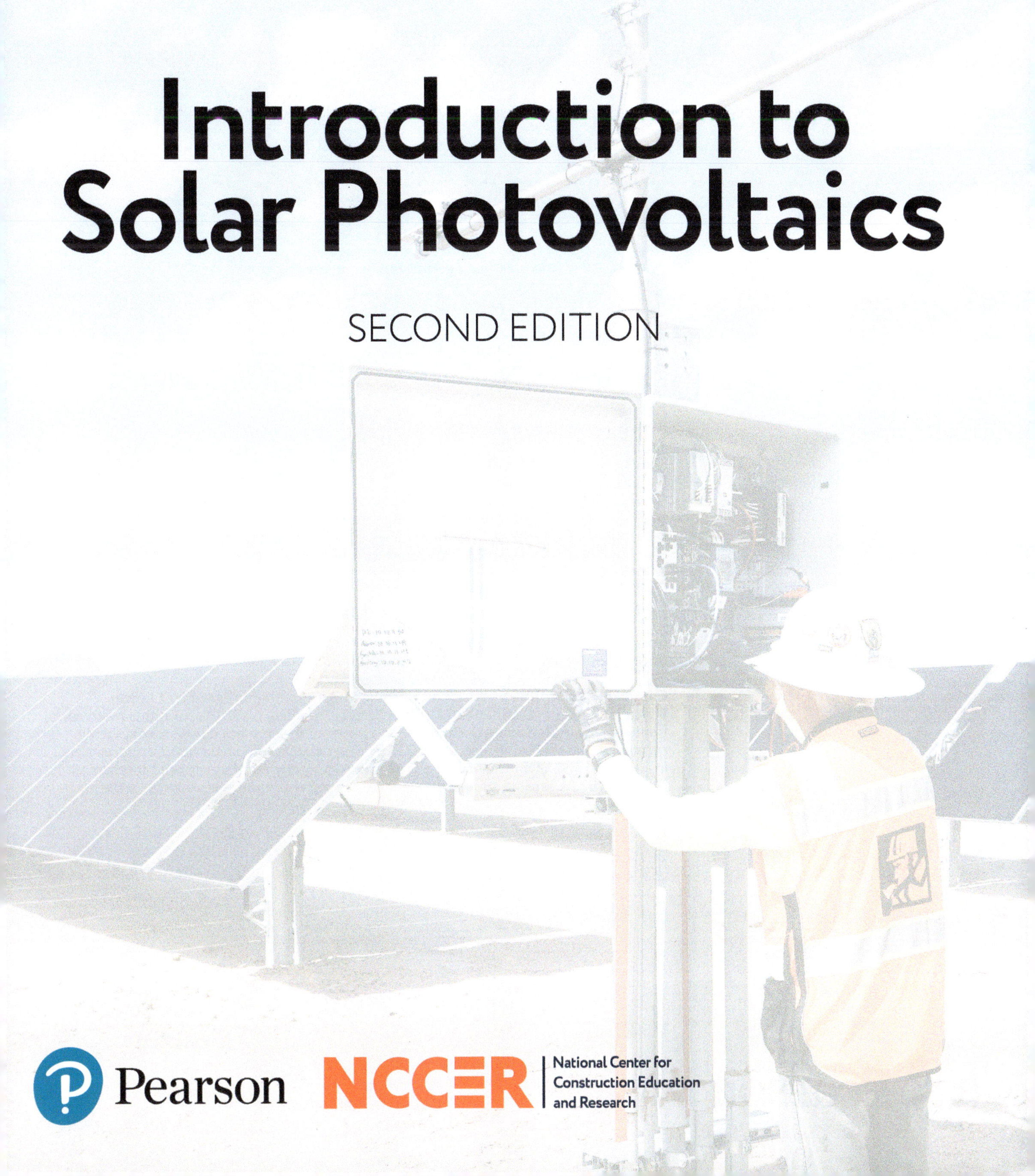

Pearson

NCCER | National Center for Construction Education and Research

NCCER
President and Chief Executive Officer: Boyd Worsham
Vice President of Innovation and Advancement: Jennifer Wilkerson
Chief Learning Officer: Lisa Strite
Senior Manager, Curriculum Development: Chris Wilson
Solar Photovoltaics Project Manager: Douglas Hancock
Technical Writing Manager: Gary Ferguson
Technical Writers: Stephanie Warnick, Karen Serink
Art Manager: Bree Rodriguez
Technical Illustrators: Judd Ivines
Production Artist: Chris Kersten
Permissions Specialist: Adam Black

Pearson
Program Manager: Vanessa Price
Senior Digital Content Producer: Shannon Stanton
Content Producer: Alexandrina Wolf, Alma Dabral
Development Editor: Nancy Lamm
Employability Solutions Coordinator: Monica Perez
Cover Designer: Mary Siener
Rights and Permissions: Jenell Forschler
Composition: Integra Software Services

Cover Image
Cover photo provided by McCarthy Building Companies, Inc.

10 9 8 7 6 5 4 3 2 1

21 2026

Pearson

NCCER National Center for Construction Education and Research

ISBN-13: 978-0-13-818595-4

PREFACE

To the Trainee

Welcome to the NCCER Solar Photovoltaic (PV) craft training curriculum! Solar power is rapidly reshaping the way energy is generated across the globe. As demand for clean energy increases, solar energy is quickly becoming an integral part of residential and industrial/commercial construction. Utility companies are also installing large solar energy centers across the country. For skilled craftworkers who want to take advantage of these new opportunities in the solar power industry, this curriculum is designed for you.

Because of governmental regulation and private initiatives, the increasing demand for qualified solar PV installers is creating many opportunities to join an expanding and viable workforce. According to the US Bureau of Labor Statistics, the increased demand is expected to grow at a much faster pace than the average occupational growth. This expectation is echoed by the US Energy Information Administration, which projects that over 900 billion kilowatt-hours of energy will be produced by solar by the year 2050.

This curriculum teaches information about specific equipment, as well as installation techniques that are valuable to solar PV systems installers. It also highlights solar PV topics, common tools, and trade-specific techniques related to site analysis, design considerations, system components, troubleshooting, and maintenance. Another advantage of this curriculum is that it helps prepare solar PV trainees for the North American Board of Certified Energy Practitioners (NABCEP) PV Associate Exam. After completing this curriculum, trainees should have the skills needed to install, maintain, and repair solar PV components and equipment on residential, commercial, and utility-scale jobsites.

New with *Solar Photovoltaics*

This second edition of *Introduction to Solar Photovoltaics* has been revised to provide modernized curriculum content and expanded to include information about residential, commercial/industrial, and utility-scale solar PV installations. Ground- and roof-mounted installations are covered, including differences in low-slope and steep-slope installation.

We wish you success as you progress through this training program. If you have any comments on how NCCER might improve upon this textbook, please complete the User Update form using the QR code. NCCER appreciates and welcomes its customers' feedback. You may submit yours by emailing **support@nccer.org**. When doing so, please identify feedback on this title by listing *#INTROTOSOLAR* in the subject line.

Our website, **www.nccer.org**, has information on the latest product releases and training.

SCAN ME

NCCER Standardized Curricula

NCCER is a not-for-profit 501(c)(3) education foundation established in 1996 by the world's largest and most progressive construction companies and national construction associations. It was founded to address the severe workforce shortage facing the industry and to develop a standardized training process and curricula. Today, NCCER is supported by hundreds of leading construction and maintenance companies, manufacturers, and national associations. The NCCER Standardized Curricula was developed by NCCER in partnership with Pearson, the world's largest educational publisher.

Some features of the NCCER Standardized Curricula are as follows:

- An industry-proven record of success
- Curricula developed by the industry, for the industry
- National standardization providing portability of learned job skills and educational credits

- Compliance with the Office of Apprenticeship requirements for related classroom training (*CFR 29:29*)
- Well-illustrated, up-to-date, and practical information

NCCER also maintains the NCCER Registry, which provides transcripts, certificates, and wallet cards to individuals who have successfully completed a level of training within a craft in NCCER's Curricula. *Training programs must be delivered by an NCCER Accredited Training Sponsor in order to receive these credentials.*

For information on NCCER's credentials and the NCCER Registry, contact NCCER Customer Service at 1.888.622.3720 or visit **www.nccer.org**.

Digital Credentials

Show off your industry-recognized credentials online with NCCER's digital credentials!

NCCER is now providing online credentials. Transform your knowledge, skills, and achievements into digital credentials that you can share across social media platforms, send to your network, and add to your resumé. For more information, visit **www.nccer.org**.

Cover Image

McCarthy Building Companies, Inc. is the oldest privately held national construction company in the country—with more than 150 years spent collaborating with partners to solve complex building challenges on behalf of its clients. With an unrelenting focus on safety and a comprehensive quality program that spans all phases of every project, McCarthy utilizes industry-leading design phase and construction techniques combined with value-added technology to maximize outcomes. Repeatedly honored as a Best Place to Work and Healthiest Employer, McCarthy is ranked the 20th largest domestic general contractor (*Engineering News-Record*, May 2018). The company has approximately 1,700 salaried employees with offices in St. Louis; Atlanta; Collinsville, IL.; Portage, IN.; Kansas City, KS.; Phoenix; Las Vegas; Denver; Dallas, Houston; Albuquerque; and San Diego, Newport Beach, San Francisco, San Jose and Sacramento, CA. McCarthy is 100% employee owned. For more information, visit **www.mccarthy.com**.

DESIGN FEATURES

Content is organized and presenteds in a functional structure that allows trainees to access the information where they need it.

Source: LeoPatrizi/Getty Images

Objectives

Successful completion of this module prepares you to do the following:

1. Understand fundamentals of solar PV, the Sun, and electrical theory as it pertains to solar photovoltaic (PV) system basics.
   a. Describe the photovoltaic process and main electrical-related codes.
   b. Describe characteristics of a net-zero building.
   c. Explain how different PV systems operate relative to the power grid.
   d. Describe how the Sun influences the position of PV modules and arrays.
   e. Understand how Watt's law and Ohm's law are used in PV system installation.
2. Describe PV system primary components.
   a. Describe different types of mechanical components for ground-mounted and roof-mounted PV systems, including motion capabilities.
   b. Explain general similarities and differences among the three main types of solar cells.
   c. Explain how PV modules from various manufacturers can be compared for performance.
   d. Describe the purpose of an inverter.
   e. Understand the differences in battery options and requirements for an energy storage solution.
3. Identify a PV system's balance of system (BOS) components.
   a. Identify mechanical BOS components within the system.
   b. Identify electrical BOS components within the system.
4. Understand safety considerations in PV systems.
   a. Describe PPE for fall protection.
   b. Understand the potential electrical hazards in working with solar PV systems.
5. Describe the project life cycle of a solar PV installation.
   a. Understand how a utility bill helps establish customer solar energy goals.
   b. Identify key steps in performing a solar site assessment.
   c. Describe key factors involved in decision making for a solar system design.
   d. Describe the key steps in system installation.
   e. Explain what to look for in a system maintenance visual inspection.
6. Understand fundamental changes happening to the power grid in response to renewable energy additions into the system.
   a. Explain power generation and distribution trends.

Digital Resources

Resources, checklists, videos, and additional articles are available on the *Digital Resources for Solar PV* web page. For example, a site assessment checklist is available to help with documenting your findings about potential installation locations.

Digital Resources for Introduction to Solar Photovoltaics

Scan this code using the camera on your phone or mobile device to view the digital resources related to this craft.

SCAN ME

Trainees can navigate *Introduction to Solar Photovoltaics* using color coded tabs on the upper right hand corner.

The Objectives list the skills and knowledge trainees need in order to complete the module successfully.

Performance Tasks	1.0.0 Build Your Knowledge
This is a knowledge-based module. There are no Performance Tasks.	**Objective** Understand fundamentals of solar PV, the Sun, and electrical theory as it pertains to solar photovoltaic (PV) system basics. a. Describe the photovoltaic process and main electrical-related codes. b. Describe characteristics of a net-zero building. c. Explain how different PV systems operate relative to the power grid. d. Describe how the Sun influences the position of PV modules and arrays. e. Understand how Watt's law and Ohm's law are used in PV system installation.

Section Openers provide a visual organizational structure for the information. Objectives and Performance tasks are broken out for each section.

Semiconductor: A material that exhibits the properties of both a conductor and an insulator.

How does the photovoltaic process work? The Sun emits photons of energy as light that strike the PV module. Based on the angle of the light to the PV module's surface, some of that energy is absorbed and some is reflected. The optimal position is perpendicular (90°) to the Sun. The absorbed energy excites the atoms in the solar cell's **semiconductor**, releasing negatively charged particles called electrons. This process transforms the solar energy into electrical energy.

Trade Terms appear on the page adjacent to the text where they are first presented.

Step-by-step math equations help make the concepts clear and easy to grasp.

Step 1 Using a sun path chart for the jobsite location, select a sun path.
Step 2 Obtain the Sun's altitude, located on the Y-axis of the chart.
Step 3 Solve the right-triangle equation for the PV module's tilt angle.

As an example, suppose the total load current in the main line equals 20A and the voltage is 120V. Using Watt's law, how much power is generated?

$$P = I \times E$$
$$P = 20A \times 120V$$
$$P = 2400W \text{ or } 2.4kW$$

If this power is used for five hours, then how much energy is used over time?

$$2.4kW \times 5 \text{ hours} = 12kWh$$

SCAN ME

QR codes link trainees directly to digital resources that highlight current content.

Important information is highlighted, illustrated, and presented to facilitate learning.

Placement of images near the text description and details such as callouts and labels help trainees absorb information.

It is optimized along a chosen sun path within the sky, associated with time of year: summer, winter, or spring/fall.

A PV module's optimal position is for its surface to be perpendicular (90°) to the Sun (*Figure 19*). There are two methods to determine the PV module's tilt angle.

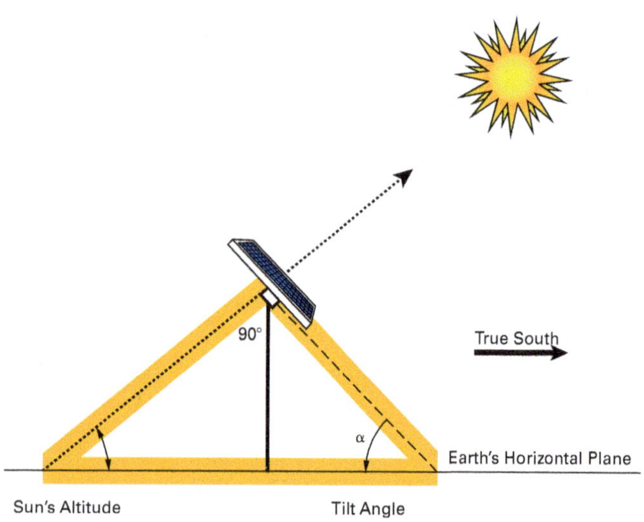

Figure 19 Adjust the tilt angle to align the solar PV module perpendicularly to the Sun.

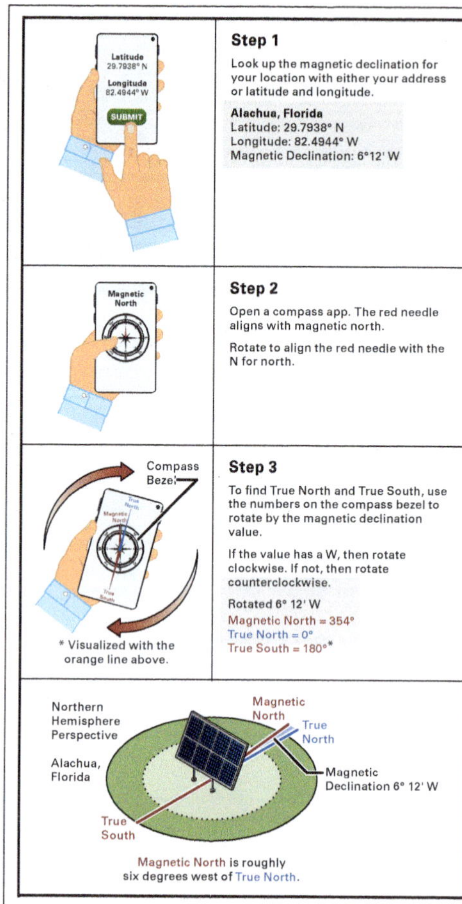

Figure 17 Magnetic north and true north for Alachua, FL.
Source: https://www.ngdc.noaa.gov/geomag/calculators/magcalc.shtml

New boxes highlight safety and other important information for trainees. Warning boxes stress potentially dangerous situations, while Caution boxes alert trainees to dangers that may cause damage to equipment. Notes boxes provide additional information on a topic.

WARNING!

If the amperage produced by the PV modules exceeds the inverter's limits and/or wire ampacity, then the wires may melt and cause a fire. Know the limits to the number of PV modules that can be strung together.

CAUTION

Do not locate the inverter or batteries more than 100 feet away from the array without adjusting the wire size. Choosing the correct wire size will limit voltage drop and is crucial for system performance and safety.

NOTE

A solar PV roof installation requires an understanding of roofing fundamentals so you do not compromise the integrity of the entire roof system. More details are provided in NCCER Module 57106, *Roofing Fundamentals* and NCCER Module 57107, *Roof-Mounted Solar PV Installation*.

Going Green

Air-Source Heat Pumps

Air-source heat pumps continue to make advancements towards decreasing a building's carbon footprint. Considered to have a 400% efficiency rating in specific environmental conditions, an air-source heat pump uses 1 kilowatt-hour (kWh) of electricity to capture 3kWh of solar thermal energy to heat or cool a building. This results in a total input of 2,500kWh of solar energy to generate 10,000kWh of electrical energy (2,500kWh × 400% = 10,000kWh). An air-source heat pump is a viable heating option for most climates.

Going Green looks at ways to preserve the environment, save energy, and make good choices regarding the health of the planet.

Did You Know?

CO_2 Emissions

In 2020, the combustion of fossil fuels to generate electricity was the second largest source of CO_2 emissions in the US, accounting for about 31% of total US CO_2 emissions and 24% of total US greenhouse gas emissions.

DC versus AC

DC flows in one direction at constant **voltage**. AC reverses direction sixty times per second with a change in volts from 0 to 120 volts (V). In the United States, residential homes use 120V or 240V, and solar PV systems may not exceed 600 VDC output at the PV modules.

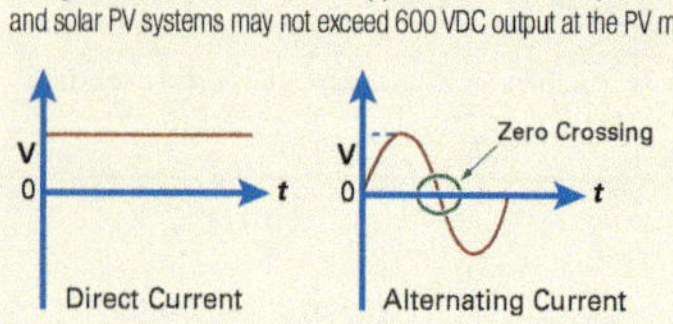

These boxed features provide additional information that enhances the text.

Think About It

Perspective

A site assessor may begin PV array sizing from the perspective of how much space is available to mount PV modules, while a system designer may begin with the customer's PV energy goals.

Review questions at the end of each section and module allow trainees to measure their progress

2.0.0 Section Review Questions

1. Which type of installation requires soil studies to guide foundational design decisions?
   a. Building-integrated system
   b. Ground-mounted system
   c. Roof-mounted system
   d. Stand-off mounted system

2. The NCCER Roofing Program defines a steep-slope roof as having a minimum rise-to-run ratio of _____.
   a. 3:12
   b. 4:12
   c. 8:12
   d. 16:12

3. A type of PV module having a characteristic flaked appearance is probably made up of _____.
   a. monocrystalline cells
   b. megacrystalline cells
   c. polycrystalline cells
   d. amorphous cells

Module 57101 Review Questions

1. The *NEC*® requirements for PV installations can be found in _____.
   a. *NEC Article 422*
   b. *NEC Article 450*
   c. *NEC Article 517*
   d. *NEC Article 690*

2. In 2020, what was the average annual electricity consumption for a US residential utility customer?
   a. 893kWh
   b. 3,600W
   c. 10,715kWh
   d. 2MW

3. Which type of system is best to use in areas where access to the power grid is inconvenient or unavailable?
   a. Grid-access system
   b. Grid-interactive system
   c. Grid-tied system
   d. Off-grid system

NCCERconnect

This interactive online course is a unique web-based supplement in the form of an electronic book that provides a range of visual, auditory, and interactive elements to enhance training. Visit **www.nccerconnect.com** for more information!

1.0.0 Basic Rigging Equipment

Regulations and Site Procedures

The information in this module is intended as a general guide. The techniques shown here are not the only methods that can be used to perform a lift. Many techniques can be safely used to rig and lift different loads.

Some of the techniques for certain kinds of rigging and lifting are spelled out in requirements issued by federal government agencies. Some will be provided at the job site, where written site procedures that address any special conditions that affect lifting procedures on that site can be found. Questions about any of these procedures should be directed to the supervisor at the site.

- They are flexible. They mold themselves to the shape of the load (*Figure 5*).
- They are very elastic, and they stretch under a load much more than wire rope. This stretching allows synthetic slings to absorb shocks and to cushion the load.
- Loads suspended in synthetic web slings are less likely to twist than those in wire rope or chain slings.

Watch: Opportunities in Construction

Source: Slingmax, Inc.

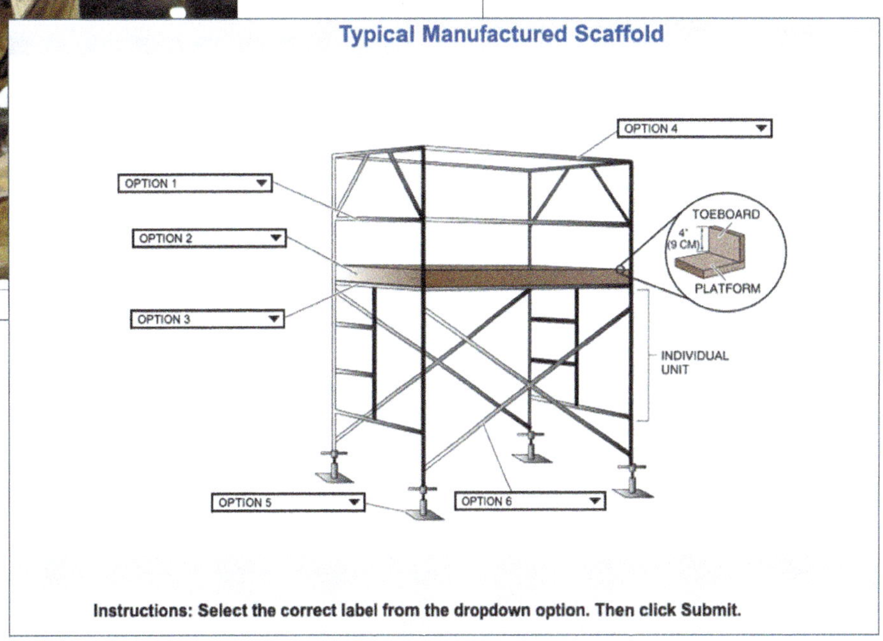

ACKNOWLEDGMENTS

This curriculum was revised as a result of the farsightedness and leadership of the following sponsors:

ABC Illinois Chapter, Inc.
Adaptive Construction Solutions
Arizona Western College
Aurora Solar
Black and Veatch
GAF
Imperial Valley College
McCarthy Building Company
Millennium Solar Electric Training Academy
Mortenson Construction
National Roofing Contractors Association
Orbital Solar Services
Pacific Northwest National Laboratory
Power52 Energy Institute
Solar Energy Industries Association
Solar Source
US Department of Energy
Yuma Private Industry Council, Inc.

This curriculum would not exist were it not for the dedication and unselfish energy of those volunteers who served on the Authoring Team. A sincere thanks is extended to the following:

Anthony Johnson
Brian Davis
Bruce Carrol
Bryan Morris
Christopher Williams
Derek Sams
Elizabeth Swiger
Franklin Brown
James Gain
Jose Roman
John Schutte
Jon Goodman
Karen Champion
Linda O'Connor
Nicholas Morgan
Philip Andries
Rick Gilbert
Robert Burns
Robert Wallace
Tiffany McMillan
Tyler Hays

NCCER PARTNERS

To see a full list of NCCER Partners, please visit **www.nccer.org/about-us/partners**.

SCAN ME

CONTENTS

Module 57101 Introduction to Solar Photovoltaics

MODULE 57101

Introduction to Solar Photovoltaics

Source: LeoPatrizi/Getty Images

Objectives

Successful completion of this module prepares you to do the following:

1. Understand fundamentals of solar PV, the Sun, and electrical theory as it pertains to solar photovoltaic (PV) system basics.
   a. Describe the photovoltaic process and main electrical-related codes.
   b. Describe characteristics of a net-zero building.
   c. Explain how different PV systems operate relative to the power grid.
   d. Describe how the Sun influences the position of PV modules and arrays.
   e. Understand how Watt's law and Ohm's law are used in PV system installation.
2. Describe PV system primary components.
   a. Describe different types of mechanical components for ground-mounted and roof-mounted PV systems, including motion capabilities.
   b. Explain general similarities and differences among the three main types of solar cells.
   c. Explain how PV modules from various manufacturers can be compared for performance.
   d. Describe the purpose of an inverter.
   e. Understand the differences in battery options and requirements for an energy storage solution.
3. Identify a PV system's balance of system (BOS) components.
   a. Identify mechanical BOS components within the system.
   b. Identify electrical BOS components within the system.
4. Understand safety considerations in PV systems.
   a. Describe PPE for fall protection.
   b. Understand the potential electrical hazards in working with solar PV systems.
5. Describe the project life cycle of a solar PV installation.
   a. Understand how a utility bill helps establish customer solar energy goals.
   b. Identify key steps in performing a solar site assessment.
   c. Describe key factors involved in decision making for a solar system design.
   d. Describe the key steps in system installation.
   e. Explain what to look for in a system maintenance visual inspection.
6. Understand fundamental changes happening to the power grid in response to renewable energy additions into the system.
   a. Explain power generation and distribution trends.

Digital Resources

Resources, checklists, videos, and additional articles are available on the *Digital Resources for Solar PV* web page. For example, a site assessment checklist is available to help with documenting your findings about potential installation locations.

Digital Resources for Introduction to Solar Photovoltaics

Scan this code using the camera on your phone or mobile device to view the digital resources related to this craft.

SCAN ME

Performance Tasks

This is a knowledge-based module. There are no performance tasks.

Overview

Solar photovoltaic (PV) modules transform solar energy, *sunlight*, into electrical energy that powers an electrical load, such as your computer or refrigerator. Using PV modules, the United States and other nations around the world can harness solar energy. Solar energy is being used to reach 2030 and 2050 global climate change initiatives. These initiatives are to reduce carbon emissions and decrease the reliance on fossil fuels. Completion of this module helps prepare solar PV trainees for the North American Board of Certified Energy Practitioners (NABCEP) PV Associates Exam and an exciting career in the solar PV industry.

NOTE

NABCEP Board Certifications and Associate Credentials are valid for three years from the date of issuance.

NCCER Industry-Recognized Credentials

If you are training through an NCCER-accredited sponsor, you may be eligible for credentials from NCCER. The ID number for this module is 57101. Note that this module may have been used in other NCCER curricula and may apply to other level completions. Contact NCCER at 1.888.622.3720 or go to **www.nccer.org** for more information.

You can also show off your industry-recognized credentials online with NCCER's digital credentials. Transform your knowledge, skills, and achievements into credentials that you can share across social media platforms, send to your network, and add to your resumé. For more information, visit **www.nccer.org**.

1.0.0 Build Your Knowledge

Performance Tasks

This is a knowledge-based module. There are no Performance Tasks.

Objective

Understand fundamentals of solar PV, the Sun, and electrical theory as it pertains to solar photovoltaic (PV) system basics.

a. Describe the photovoltaic process and main electrical-related codes.
b. Describe characteristics of a net-zero building.
c. Explain how different PV systems operate relative to the power grid.
d. Describe how the Sun influences the position of PV modules and arrays.
e. Understand how Watt's law and Ohm's law are used in PV system installation.

According to the US Energy Information Administration (EIA) projections, the use of solar power will make up 14% of total energy generation in the United States by 2035 and 20% by 2050. As a result, solar installation offers a wide range of career opportunities in a growing industry with the creation of 400,000 jobs by 2030 and 900,000 positions by 2035. Installation jobs make up more than half of all solar PV industry employment, with other job openings in manufacturing, sales, distribution, operation, and maintenance (*Figure 1*).

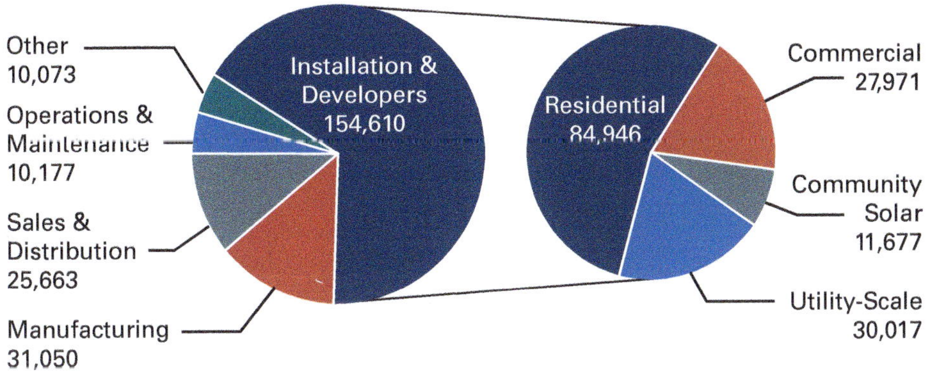

Figure 1 Solar industry job distribution.
Source: Data from National Solar Job Census 2020, Interstate Renewable Energy Council (IREC)

Let's begin by building your knowledge about the history of solar energy, some basic concepts about the Sun, and electrical theory as applied to solar photovoltaic (PV) systems. Then, you'll learn about the solar PV system components and a job's project life cycle, explaining how PV systems are sized, designed, installed, and maintained.

Solar energy is a **clean energy** and **renewable energy**. The **solar photovoltaic (PV) module** transforms solar energy into electricity that powers electrical loads to do work. The system is quiet, reliable, and often requires little maintenance. It can also be a source of income, allowing home and business owners to credit energy to the utility company's power grid. Multiple PV modules make an **array**. Add the remaining PV components and you've got a solar PV system.

Solar PV systems can be used within the following sectors: residential, commercial, industrial, and utility scale (*Figure 2*).

Clean energy: An energy source that doesn't emit greenhouse gases.

Renewable energy: An energy source that quickly replenishes the supply of energy.

Solar photovoltaic (PV) module: A PV system component consisting of numerous electrically and mechanically connected PV cells encased in a protective glass or laminate frame. Also known as a *PV panel*.

Array: Multiple solar photovoltaic (PV) modules wired together to make a larger structure.

(A) Residential Steep-Slope Roof Mount Stand-Off, Rack Mount **(B) Commercial Low-Slope Roof Mount, Ballast Mount** **(C) Utility Scale Ground Mount**

Figure 2 Industry sectors.
Source: (A) Douglas Cliff/Getty Images; (B) Asia-Pacific Images Studio/Getty Images; (C) 4kodiak/Getty Images

1.1.0 Solar Energy and Related Codes

Solar-thermal power is a form of solar energy which has been used by humans for over 2,000 years. It is the Sun's **solar irradiance** felt as heat. The Sun's heat can preserve food, heat water, dry clothes, and sustain life on planet Earth. The ancient Greeks built thick walls to trap heat during the day and release it slowly at night. They also oriented buildings to provide shading in summer while maximizing sunlight in winter. These are passive forms of solar power because the Sun's heat is used without transforming it into another form of energy.

Active forms of solar power were introduced in 1839, when Alexander Becquerel discovered that certain materials produce electric current when exposed to light. This eventually led to the invention of the first primitive PV cell.

Solar irradiance: The amount of solar radiation that strikes Earth's surface in a given span of time (kW/m^2/days).

Solar photovoltaic (PV) cell: A semiconductor device that converts the energy of sunlight into direct current. Also known as a *solar cell*.

Semiconductor: A material that exhibits the properties of both a conductor and an insulator.

By 1891, Clarence Kemp patented the first water heater powered by solar energy, but it wasn't until 1954 that Bell Laboratories modified silicon cells to invent the first practical **solar photovoltaic (PV) cell**.

The solar PV system continued to evolve, transforming solar energy to electrical energy through the photovoltaic process as shown in *Figure 3*.

How does the photovoltaic process work? The Sun emits photons of energy as light that strike the PV module. Based on the angle of the light to the PV module's surface, some of that energy is absorbed and some is reflected. The optimal position is perpendicular (90°) to the Sun. The absorbed energy excites the atoms in the solar cell's **semiconductor**, releasing negatively charged particles called electrons. This process transforms the solar energy into electrical energy.

Like a battery, PV modules have a positive terminal and a negative terminal within the sealed junction box at the back of the module. The junction box is also called the *termination box*. The termination box is normally prewired with positive and negative leads. These leads are called *panel wiring*.

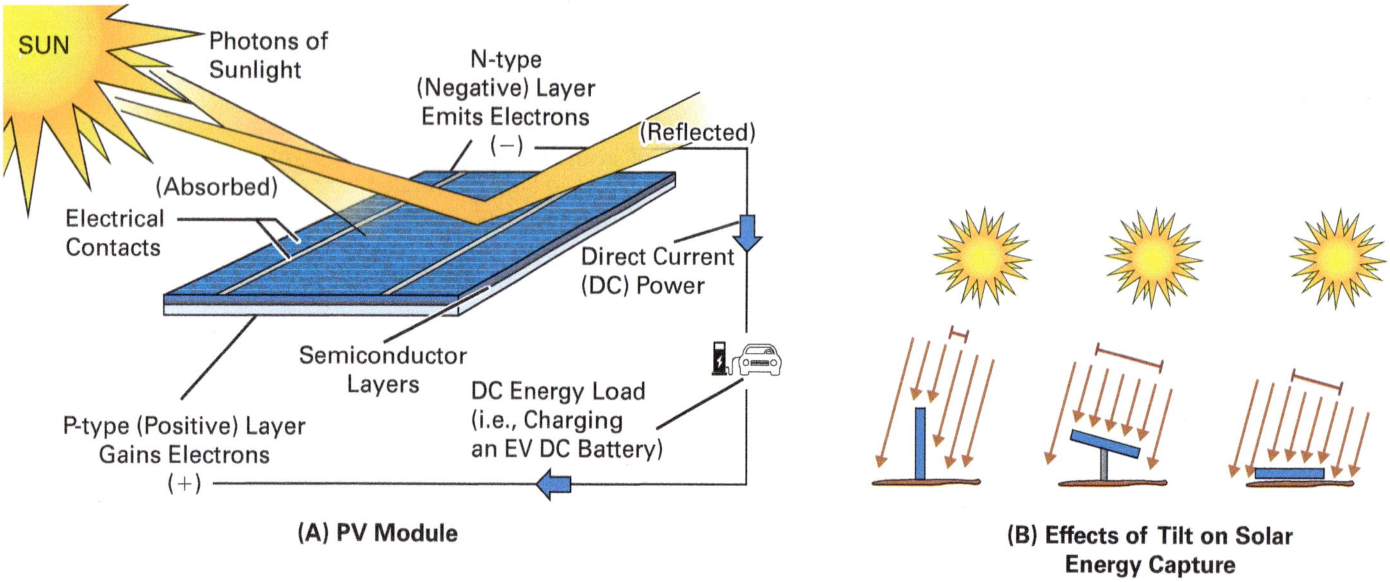

Figure 3 Photovoltaic process.

Balance of system (BOS): The additional parts which connect the primary components together and are required to execute a solar PV installation based on the engineer's solar plan.

The negatively charged electrons produce a direct current (DC), which is drawn to the positive terminal and into the circuit towards the electrical load(s).

The electricity travels along the electrical circuit through the **balance of system (BOS)** components, which include wires, fuses, and disconnect switches. These components control the flow of electricity through the solar PV system's primary components, such as the inverter and optional energy storage. During this process, the energy is inverted to alternating current (AC), supplying power to the electrical load. Most appliance motors require AC power to operate.

People continue to invent new applications for solar power, like outdoor string lighting and construction signs. They are also implementing space efficiency

Source: Naeblys/Getty Images

Did You Know?

Advancing Space Exploration

Solar energy is still the primary source of power in satellite systems and spacecraft. NASA's Lucy spacecraft launched in 2021 on its 12-year mission to the Jupiter Trojan asteroids, as shown in this illustration by Southwest Research Institute. The solar wings span a five-story building. The solar PV modules generate around 500 watts. This isn't a lot of energy, about enough to run a washing machine, but Lucy will be far away from the Sun, so the wings need to be large to take in as much solar energy as possible.

techniques, such as agrivoltaics, which creates two sources of income for the same acreage of farmland. Solar carports are another example of space efficiency.

Going Green

Air-Source Heat Pumps

Air-source heat pumps continue to make advancements towards decreasing a building's carbon footprint. Considered to have a 400% efficiency rating in specific environmental conditions, an air-source heat pump uses 1 kilowatt-hour (kWh) of electricity to capture 3kWh of solar thermal energy to heat or cool a building. This results in a total input of 2,500kWh of solar energy to generate 10,000kWh of electrical energy (2,500kWh × 400% = 10,000kWh). An air-source heat pump is a viable heating option for most climates.

Did You Know?

US Policy Driving the Industry

Solar energy is a renewable energy. In response to climate change, the United States government has phased initiatives to use 100% renewable energy by 2050. These initiatives align with the international climate change response and goals outlined in the United Nations' (UN) Paris Agreement.

The following policies influence development around the world to reduce carbon emissions by switching to clean energies:

- *UN Paris Agreement*
- *The Long-Term Strategy of the United States*

1.1.1 Codes and Standards

All the applications for solar power are associated with construction industry codes and standards. Building codes provide minimum standards to guard the life, health, and safety of the public by regulating and controlling the design and construction of buildings, and the quality of materials used in construction. The solar photovoltaic (PV) industry uses many local and national building codes, including standards from different crafts, such as carpentry, roofing, and electrical. The following organizations provide solar industry standards:

- American National Standards Institute (ANSI)
- ASTM International
- Institute of Electrical and Electronics Engineers® (IEEE®): *IEEE 1547*
- International Code Council (ICC)
- International Building Code® (IBC®): *Section 3111 – Solar Energy Systems*
- International Residential Code® (IRC®): *R324 – Solar Energy Systems*
- International Solar Energy Provisions® (ISEP®): *Section RS402 (R324) – Solar Energy Systems*
- International Association of Plumbing and Mechanical Officials® (IAPMO®): *IAPMO IS 34*
- National Fire Protection Association® (NFPA®): *(NFPA 70®) National Electrical Code® (NEC®* and *Standard for Electrical Safety in the Workplace® (NFPA 70E®))*
- Occupational Safety and Health Administration (OSHA)
- Underwriters Laboratories® (UL®)

Some important codes are shown in *Figure 4*.

NEC Article 690 addresses PV systems other than the PV generating plant (solar farms) covered in *NEC Article 691*. The *NEC Article 705 Interconnected Electric Power Production Sources* covers the location where the building and the power grid are connected to one another. Note that the *NEC®* articles referenced within this module are from the *NEC®* 2023 code book and that US states' adoption of the *NEC®* guidelines vary; follow guidelines for your specific location.

Underwriters Laboratories® (UL®) provides product safety certification, testing, and inspection services. Some examples are *UL Standard 1703: UL Standard for Safety, Flat-Plate Photovoltaic Modules and Panels* and *UL Standard 1741: Standard for Inverters, Converters, Controllers and Interconnection System Equipment for Use with Distributed Energy Resources.*

1.2.0 Net-Zero Buildings

The first step toward *going green* starts with making buildings energy efficient. **Energy efficiency** means reducing a building's energy consumption by using appliances and lighting solutions designed to use less energy than other available products on the market.

Energy efficiency: Reducing a building's energy consumption by using appliances and lighting solutions that are designed to efficiently use energy.

Once energy efficiency measures are implemented, then the solar PV system can be sized to generate energy for the building's new level of energy consumption.

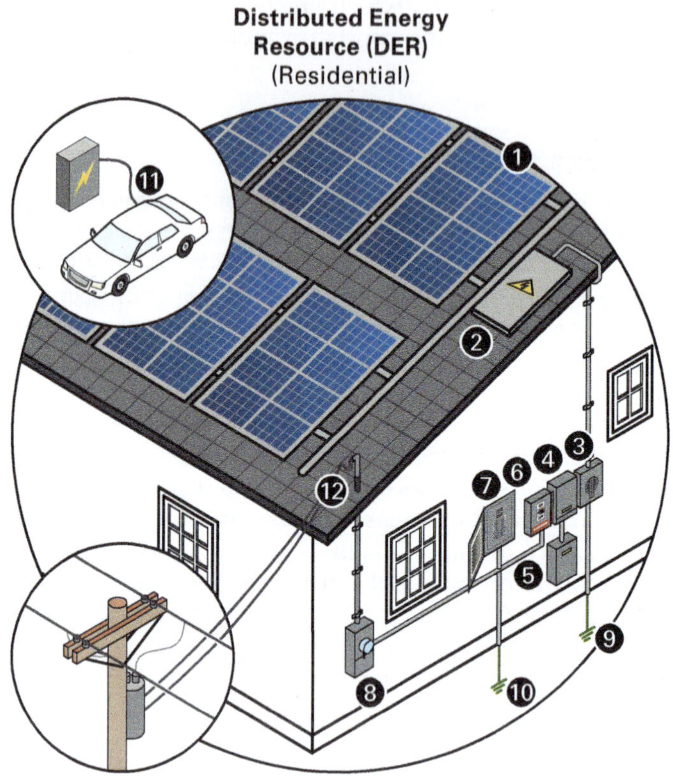

1. PV Modules
   MC4 Connector
   NEC Article 690.33
2. Combiner Box
   NEC Article 690.9
3. DC Disconnect
   NEC Article 690.13
4. Smart Inverter
   IEEE 1547 Interconnection
   UL 1741
5. Energy Storage System (ESS)
   NEC Article 706
6. AC Disconnect
   NEC Article 690.13
7. Panelboards and
   Overcurrent Protection
   NEC Articles 408 and 240
8. Point of Interconnection and
   Metering
   NEC Articles 705 and 230
9. DC Grounding
   NEC Article 690.41
10. Grounding Electrode System
    (AC Grounding)
    NEC Article 250
11. EV Power Transfer System
    NEC Article 625
12. Service
    NEC Article 230

Figure 4 *NEC*® and solar PV.

Implementing energy efficiency measures and using renewable energies—like solar PV—can create a **net-zero building**. A net-zero building generates and consumes the same amount of energy within a year, resulting in a net use of zero.

Net-zero building: A building that generates and consumes the same amount of energy within a year, resulting in a net use of zero.

According to the US Energy Information Administration, in 2020, the average annual electricity consumption for a US residential utility customer was 10,715 kilowatt-hours (kWh), an average of about 893kWh per month. For that home to be net-zero, it would have to annually produce 10,715kWh to offset the energy used. This concept applies to any building.

Customers often receive government incentives and utility rebates when implementing energy efficiency measures or a solar PV system.

Table 1 lists typical solar PV system sizes by sector.

TABLE 1 Standard System Sizes in kW DC

PV Sector	Size Range
Residential	3–10kW
Commercial/Industrial	10kW–2MW
Utility Scale	>2MW

Source: U.S. Solar Photovoltaic System Cost Benchmark: Q1 2018, National Renewable Energy Laboratory (NREL)

1.2.1 Net-Zero Emissions

Net-zero emissions is also a going green topic. A net-zero emissions building uses renewable energy sources to power electrical loads. Eliminating dependencies on fossil fuels (e.g., coal, natural gas, and oil) reduces greenhouse gas emissions. A US decarbonization goal is to minimize fossil fuel emissions to slow global warming.

Global warming is the increase in Earth's surface temperature over time. Earth's average surface temperature has increased by roughly 0.8° Celsius (C) during the past 40 years. Carbon dioxide (CO_2), which is considered a greenhouse gas, is believed to contribute to this trend because within Earth's atmosphere, it can trap heat generated by human activities and by the Sun. CO_2 is a by-product of fossil fuel combustion.

Building owners can work toward net-zero emissions goals by reducing greenhouse gases through electrification, which is the replacement of fossil-fuel based appliances with efficient electric ones. Examples of upgrades include the following:

- Switching gas water heaters to electric, heat-pump water heaters
- Using induction stoves
- Replacing oil, natural gas, and propane furnaces with air-source heat pumps

New construction practices are leading to innovation with decarbonization solutions planned from the start. *Figure 5* shows a few ideas for making an energy efficient, net-zero building.

Did You Know?

CO_2 Emissions

In 2020, the combustion of fossil fuels to generate electricity was the second largest source of CO_2 emissions in the US, accounting for about 31% of total US CO_2 emissions and 24% of total US greenhouse gas emissions.

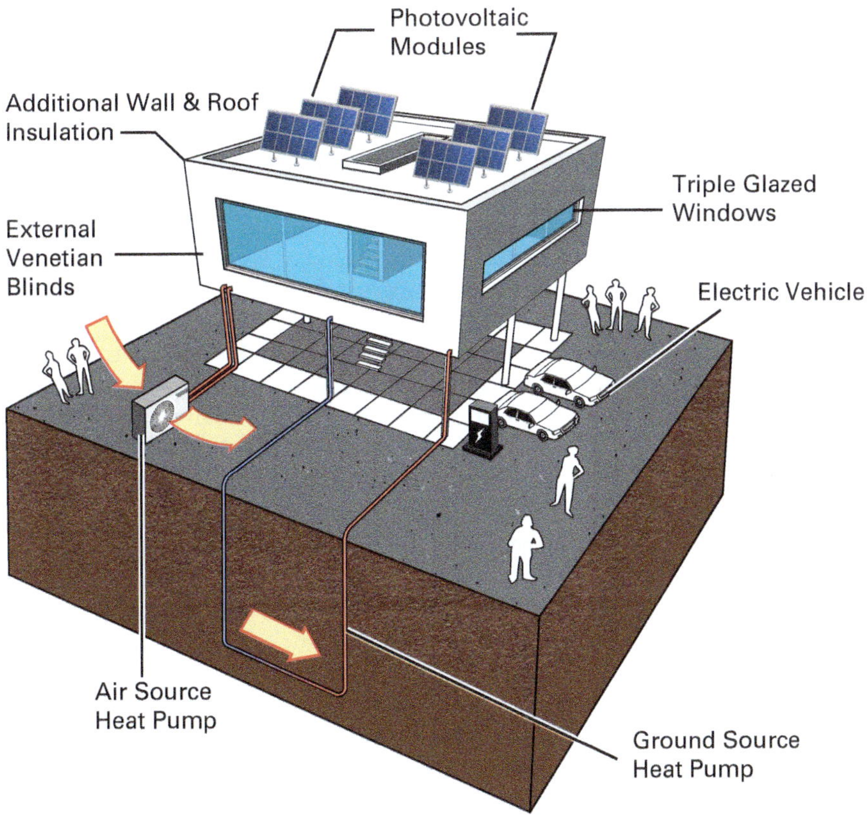

Figure 5 Net-zero emissions buildings of the future.
Source: valigursky/Getty Images

1.3.0 Solar Photovoltaic Applications

Solar PV systems can be found on rooftops, in open fields and farmland, on watches and electronics, and on satellites traveling in distant parts of the solar system.

When installing PV systems on residential and commercial buildings, an important part of the installation is site selection. When choosing an installation site, a site assessor tries to maximize the use of space available on a jobsite, whether that space is available on the ground or on a building's rooftop. It is also important to consider the PV system's relation to the power grid. The three basic types of solar PV systems are:

Off-grid system: A solar PV system with an energy storage solution that is independent of the power grid. Also known as a *stand-alone system*.

Grid-tied system: A solar PV system that operates in parallel with the power grid, providing supplemental power to the building but unable to be autonomous from the system.

Grid-interactive system: A grid-tied system with an energy storage solution.

Hybrid power system: A solar PV system with multiple renewable energy sources, such as hydro power, solar photovoltaics, solar thermal, wind turbines, or generators.

- **Off-grid system** — The solar PV system is not connected to the utility power grid and may include a battery for energy storage or a backup power generator.
- **Grid-tied system** — The solar PV system is connected to the utility power grid but lacks energy storage.
- **Grid-interactive system** — The solar PV system is a grid-tied system with energy storage.

When a solar PV system includes additional types of energy sources, such as wind turbines or generators, it is known as a **hybrid power system**.

1.3.1 Off-Grid Systems

Off-grid PV system applications are commonly used to provide power to devices, equipment, or buildings in areas where the power grid is inaccessible. For this reason, they are known as being *stand-alone*. These systems use the following two ways to deliver power:

- *Direct-drive* — operating without batteries
- *Battery-backup storage* — using batteries at night or at other times when there is no solar access

Figure 6 Direct-drive system running a water pump.
Source: AVIJIT BAITALIK/Getty Images

Direct-drive systems power DC loads, such as ventilation fans, irrigation pumps (*Figure 6*), and remote cattle watering systems. Because these systems operate only when the Sun is shining, it is essential to match the power output of the system to the anticipated load.

Battery-backup systems have a PV module that transforms solar energy into electricity to charge one or more batteries. The battery system then supplies the device or building with power when sunlight is not available (e.g., night, bad weather). A **direct current (DC)** load or an **alternating current (AC)** load may be supported. However, the AC load requires an **inverter** to convert the DC to AC power. Battery-backup systems come in many forms, including handheld electronics, streetlights, emergency highway signage, and portable generators used to provide emergency power and remote data monitoring (*Figure 7*).

Direct current (DC): An electrical current that flows in one direction.

Alternating current (AC): An electrical current in which the direction of electron flow switches back and forth in cycles.

Inverter: A device used to convert direct current to alternating current.

Autonomy: The number of days a fully charged battery system can supply power to a building's critical loads without recharging.

For many off-grid systems, when sunlight isn't available and the battery power is depleted, backup power is provided by an engine-driven generator or wind turbine. **Autonomy** is the number of days a fully charged battery system can supply power to loads without recharging. Off-grid systems are typically designed to provide three days of autonomy. This value is used when sizing the battery bank to power critical loads, also known as *essential loads*.

One possible configuration of common solar PV components is shown in *Figure* 8. Note the inclusion of the generator.

Each component has a specific purpose (*Table* 2) that will be covered in this module. Note that this is not an exhaustive list of components or combination of components.

Recent trends in the solar industry are to combine features formerly handled by multiple devices into a single device. Combining devices reduces system issues caused by mixing products from various manufacturers and also reduces the labor required for installation.

(A) Street Lights **(B) Construction Signs** **(C) Power Generators**

Figure 7 Battery-backup systems.
Source: (A) hejuan/Getty Images; (B) Jens_Lambert_Photography/Getty Images; (C) Alex Ramsay/Alamy Stock Photo

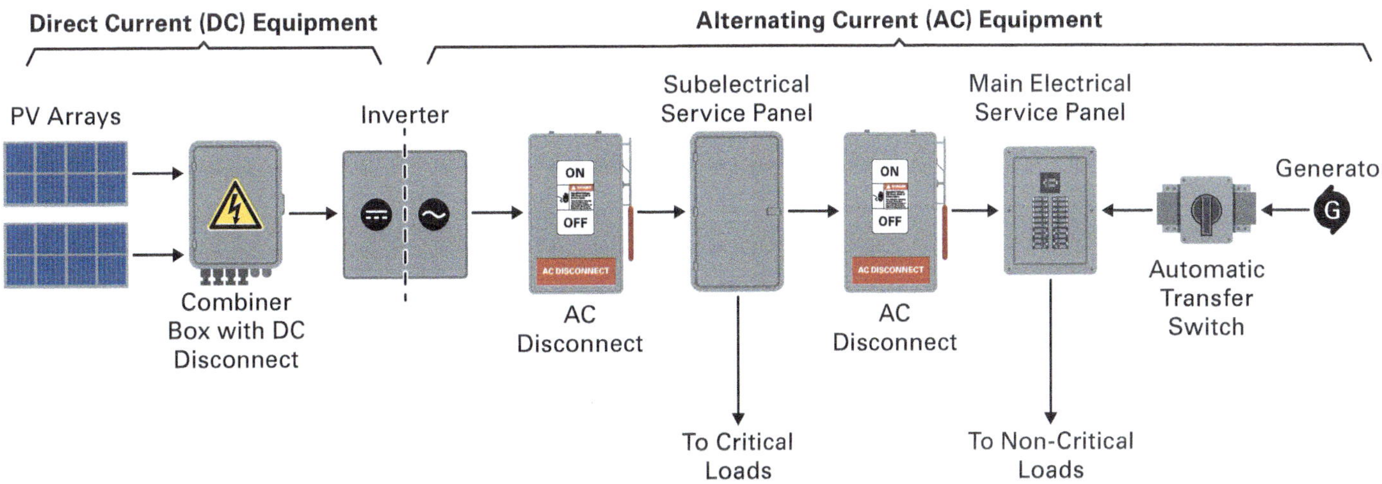

Figure 8 Off-grid configuration.

1.3.2 Grid-Tied Systems

Grid-tied systems are connected to the utility company's power grid at the building's electrical meter, or *utility meter*. Whether the meter is an older, standard meter or a newer *smart* meter, electricity can flow in both directions from the building to the power grid and from the power grid to the building. A smart meter is an electrical device that remotely and securely communicates information to the power grid's central computer system, such as the building's electric energy consumption. This is a point of interconnection.

Did You Know?

Enough Energy for All

Earth receives more energy from the Sun in a single hour than the entire world uses in a whole year.

DC versus AC

DC flows in one direction at constant **voltage**. AC reverses direction sixty times per second with a change in volts from 0 to 120 volts (V). In the United States, residential homes use 120V or 240V, and solar PV systems may not exceed 600 VDC output at the PV modules.

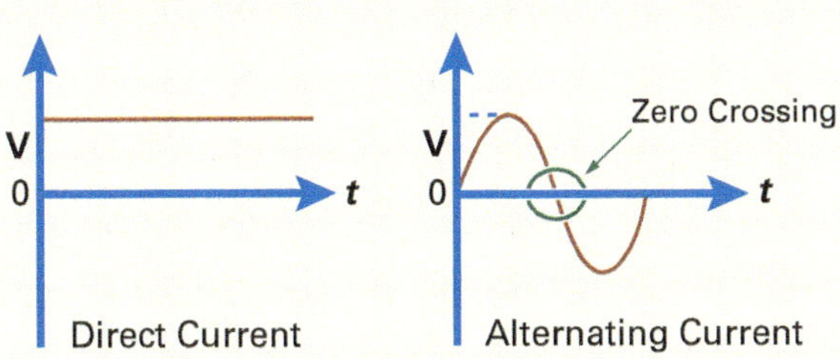

Voltage: The electric pressure between two points, originating at an electrical circuit's power source, which pushes electrons through a circuit.

Combiner box: An electrical distribution box with DC safety overprotection used to combine multiple sets of strings of PV modules into one set of conductors via a fused terminal block.

Charge controller: A device used to regulate the charging and discharging of a battery system to prevent overcharge and excess discharge.

TABLE 2 PV Components' Purpose

Component	Purpose
PV module	Generates DC electricity from solar energy
Combiner box	Combines multiple sets of strings of PV modules into one set of conductors via a fused terminal block
DC disconnect	Safety: isolates the PV array's power from the circuit
Charge controller	Regulates the electricity flow to and from the battery
Energy storage	Provides multiple batteries to store energy
Inverter	Converts DC to AC; AC is required for most appliances
Generator	Gas-powered, engine-driven power generator for emergencies
Subpanel	Connects all critical or essential loads, like the refrigerator; provides power to these types of appliances first
Load	An electrical component, such as lights or appliances, drawing power to do work
Main electrical panel	Distributes electricity to loads, includes safety features, and connects with the utility meter
Non-critical loads	Electricity required for appliances that aren't life-critical
AC disconnect	Safety: isolates the solar system from the building
Grounding system	Safety: excess electricity in the circuit travels along this wire to Earth for safe discharge
Wires	Conductors with insulation that carry electricity
Raceway	Path that protects wires from the elements or from accidentally being cut

North American Board of Certified Energy Practitioners (NABCEP): A volunteer board of renewable energy system professionals that provides standardized testing and certification for PV system installers.

Interconnection: The technical and procedural matters associated with operating interactive PV systems and other distributed generation sources in parallel with the electric utility system, as defined by NABCEP.

Net metering: A utility program enabling customers to credit their solar energy surplus into the power grid to draw upon later.

The **North American Board of Certified Energy Practitioners (NABCEP)** defines the term **interconnection** as "the technical and procedural matters associated with operating interactive PV systems and other distributed generation sources in parallel with the electric utility system."

Procedural matters include items such as permitting and agreements between the utility company and the building owner to credit the building owner's excess energy into the utility company's power grid. At night and during periods when the building's energy demands exceed the solar PV system's output, electricity is drawn from the utility company's power grid. This utility program is called **net metering**.

During a power outage or emergency, the utility company disconnects the building from the power grid via the smart meter, which prevents back-feeding power into the grid and endangering the lives of utility workers. Without an energy storage solution, the building also experiences the power outage.

Figure 9 shows the configuration of common solar PV components in a grid-tied solar PV system for powering a building. Note the use of optimizers at the PV modules.

1.3.3 Grid-Interactive Systems

A grid-interactive system is a grid-tied system with an energy storage solution (ESS), such as a bank of batteries. These batteries supply power to critical loads during power outages and during times, such as at night or during bad weather, when PV modules don't generate electricity.

Figure 10 shows the configuration of common solar PV components in a grid-interactive solar PV system for powering a building.

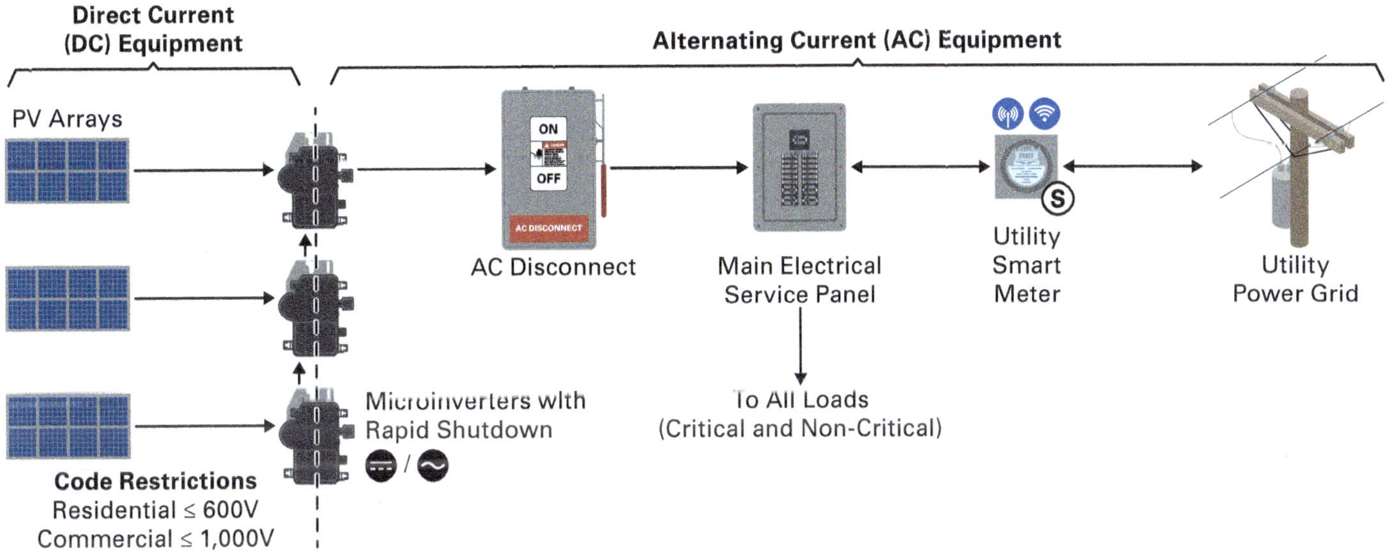

Figure 9 Grid-tied configuration.

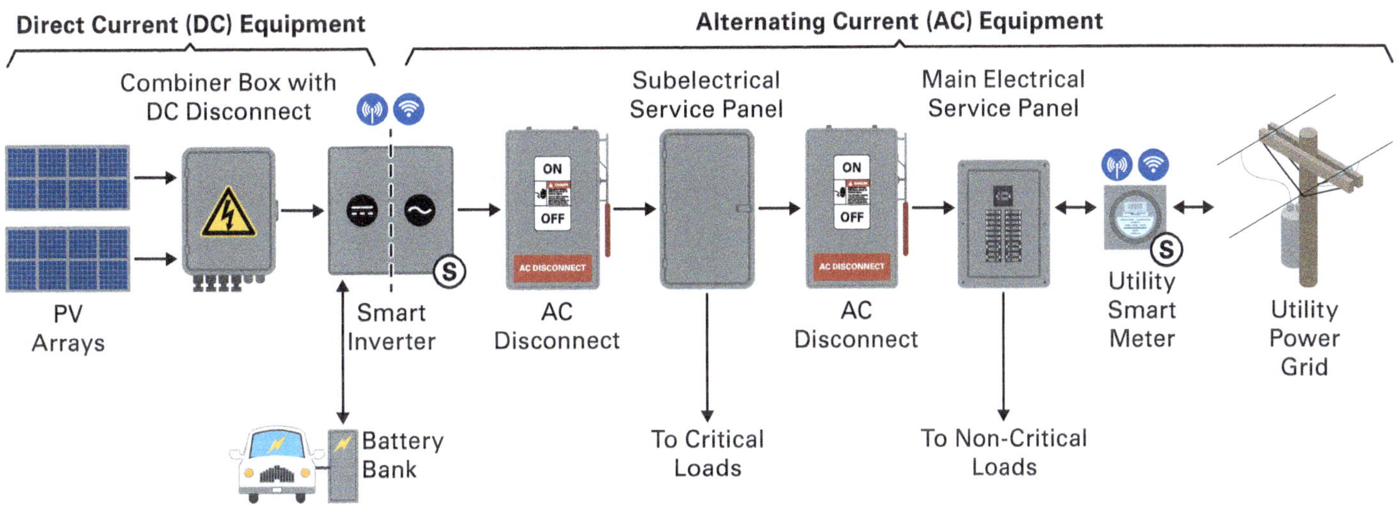

Figure 10 Grid-interactive configuration.

1.3.4 Utility Scale

To safely manage the volume of electricity in utility-scale applications, additional equipment is used (*Table 3*). *Figure 11* shows one possible system design. Grounding occurs throughout the system.

TABLE 3 Utility Scale: Additional PV Components' Purpose

Component	Purpose
Transformers (LV/MV)	Steps up low voltage to medium voltage
Switchgear	Safety: overprotection and safety disconnect
Transformers (MV/HV)	Steps up medium voltage to high voltage
Batteries	High-capacity energy storage
Grounding grid	Network of conductors placed in the soil beneath the substation, bonded to all metal noncurrent-carrying equipment and structures, such as equipment housings and fences

Inverters, transformers, and switchgear components can be *containerized*, meaning that they are bundled together into a container that is manufactured off-site and delivered to the site for installation. This technique enables faster installations, includes climate control, and provides safety factors like overprotection and safety disconnects.

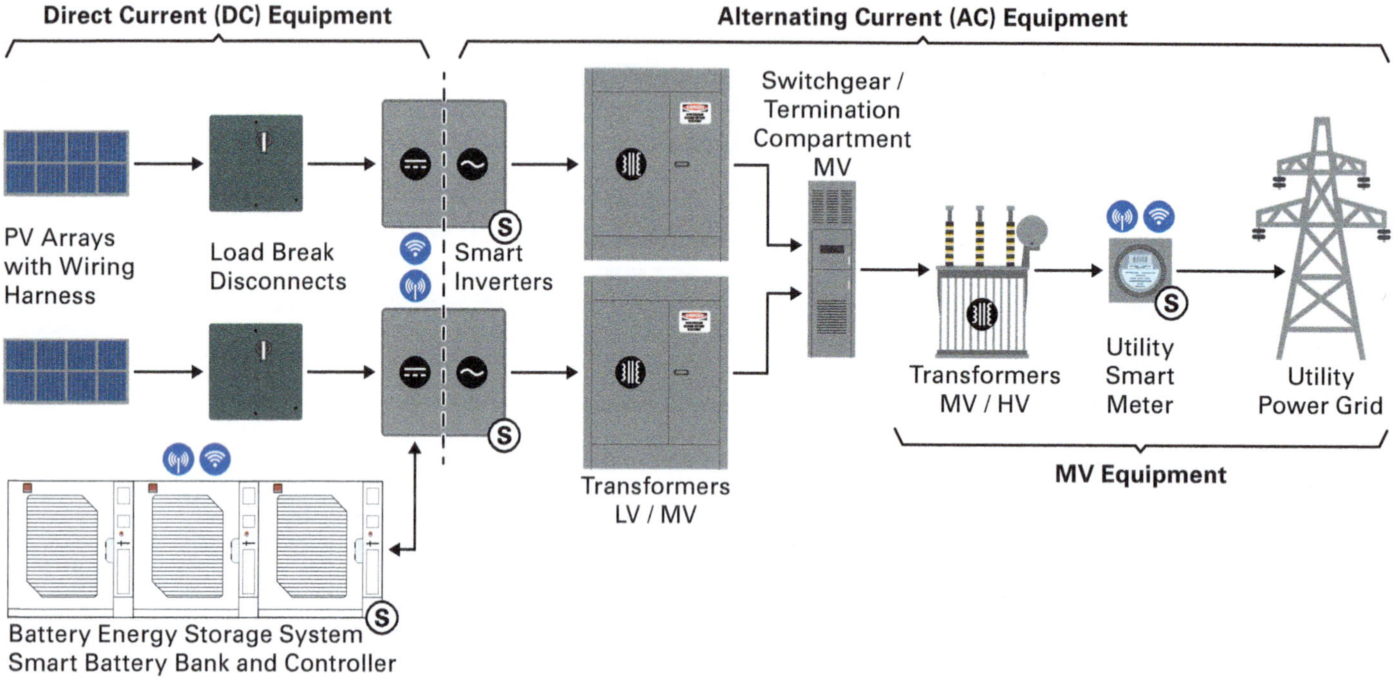

Figure 11 Potential utility-scale solar PV design.

1.4.0 The Sun

Knowing some basic information about the Sun will help you to:

- Identify potential solar PV installation locations at a jobsite during a site assessment.
- Optimize the energy yield from the PV modules by knowing how to align the PV modules' surface with the Sun.

Remember that Earth does the following:

- Tilts on its axis
- Spins on that axis, completing one rotation during a 24-hour period, or day
- Moves in an elliptical pattern, or *orbit*, around the Sun, completing the cycle in one calendar year

These characteristics of Earth's movement in the solar system—and its relation to the Sun—affect how much light reaches a PV module throughout the year. As shown in *Figure 12*, the Sun casts shadows on Earth, creating night and day. Earth's tilt and where Earth is located along its orbit creates seasonal differences. For example, in the Northern Hemisphere, the shadows are longer in winter because the Sun appears lower in Earth's sky during this time of year.

Objects on Earth's surface also cast shadows when struck by sunlight. For example, the shadows from trees and skyscrapers obstruct the PV module's solar access and potentially decrease the performance of the solar PV array. Shading is considered during project planning.

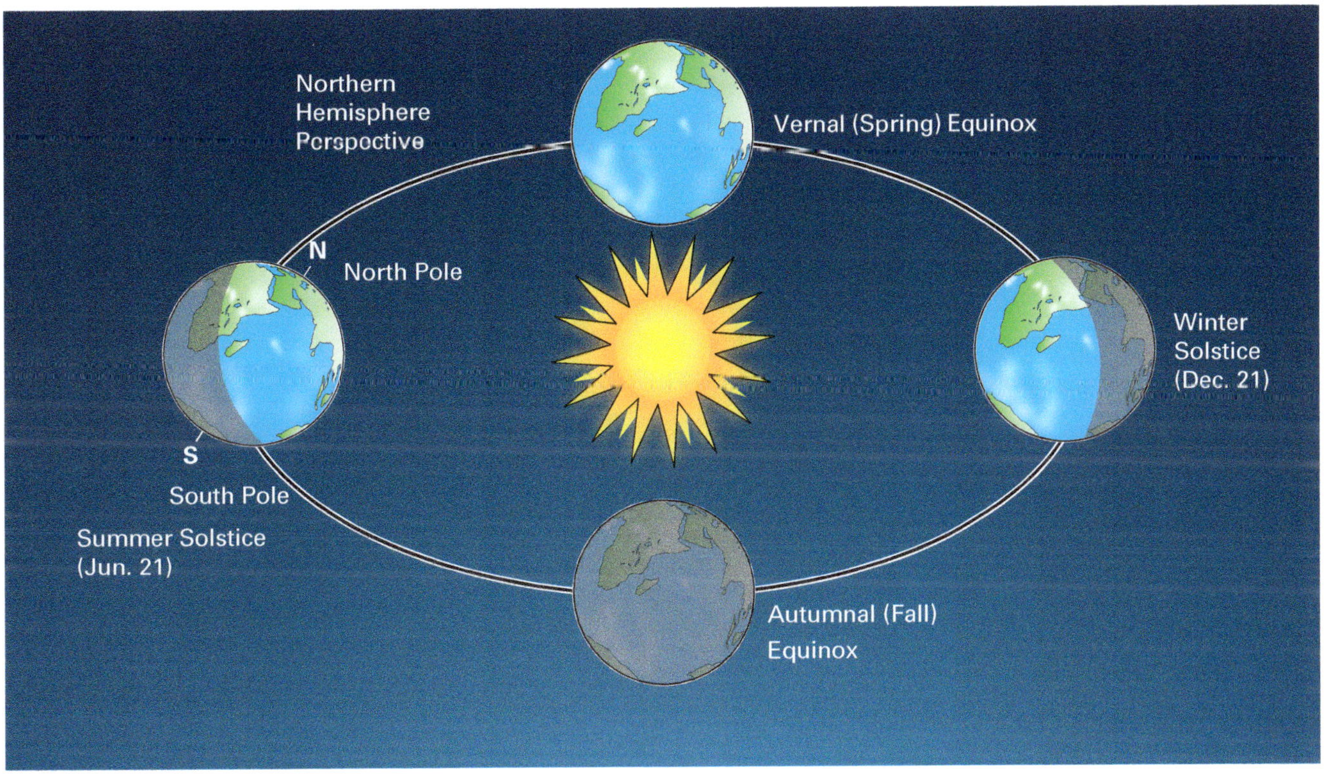

Figure 12 Earth's movement in the solar system.

1.4.1 The Sun's Path in Earth's Sky

When installing a solar PV system, the goal is to align it with the Sun. This means it is important to determine the location of the Sun in the sky.

When you look up into the sky, you observe the Sun rising in the east and setting in the west. If you were to take time-lapse photos of a jobsite in the Northern Hemisphere throughout the year, you would observe the following characteristics of the Sun's path across Earth's sky:

- Higher in the sky in summer (June)
- Lower in the sky in winter (December)
- Favors a southern trajectory across the sky

In the Southern Hemisphere, the Sun favors a northern trajectory, summer happens in December, and winter happens in June.

The summer solstice and winter solstice represent the longest and shortest days of the year, respectively, and are used to optimize the position of the PV modules.

Optimizing for the summer solstice gathers the maximum amount of sunlight within the year, which also results in the highest energy yield, but optimizing with the winter solstice may be beneficial for off-grid systems to help them get the highest yield on potentially the worst day in the year (i.e., fewer hours of sunlight, bad weather).

In *Figure 13*, you can visualize the Sun's movement as a *solar window*. This area represents the optimal positioning of the PV modules in relation to the Sun. Notice Earth's horizon is visualized as a 360-degree circle with north at 0° and south at 180°.

1.4.2 Sun Path Charts

The Sun's movement can be plotted on a **sun path** chart (*Figure 14*), which helps installers at a particular location visualize the Sun's position in the sky throughout the year. The Y-axis of the chart is the Sun's **elevation**, or **altitude**, in Earth's sky in relation to Earth's horizon. The X-axis represents the azimuth, or position of the Sun along Earth's horizon relative to north (0° azimuth). South is found at 180° azimuth, and both altitude and azimuth are measured in degrees.

Sun path: The observed movement of the Sun across Earth's sky.

Elevation: The height of something in reference to a given point.

Altitude: In relation to solar PV, the Sun's elevation in Earth's sky in relation to Earth's horizon, measured in degrees.

NOTE

When the terms *north* and *south* are used with a sun path chart, they refer to true north, also known as the North Pole, and true south, also known as the South Pole.

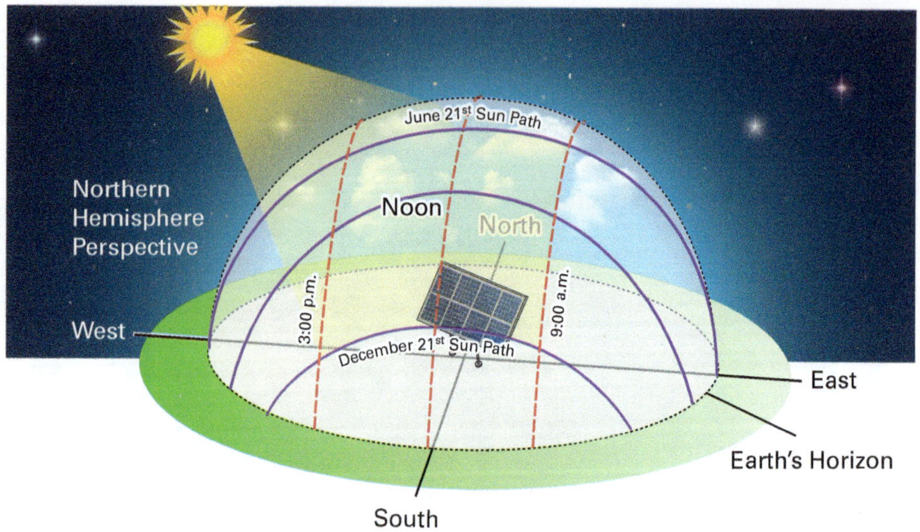

Figure 13 The solar window.

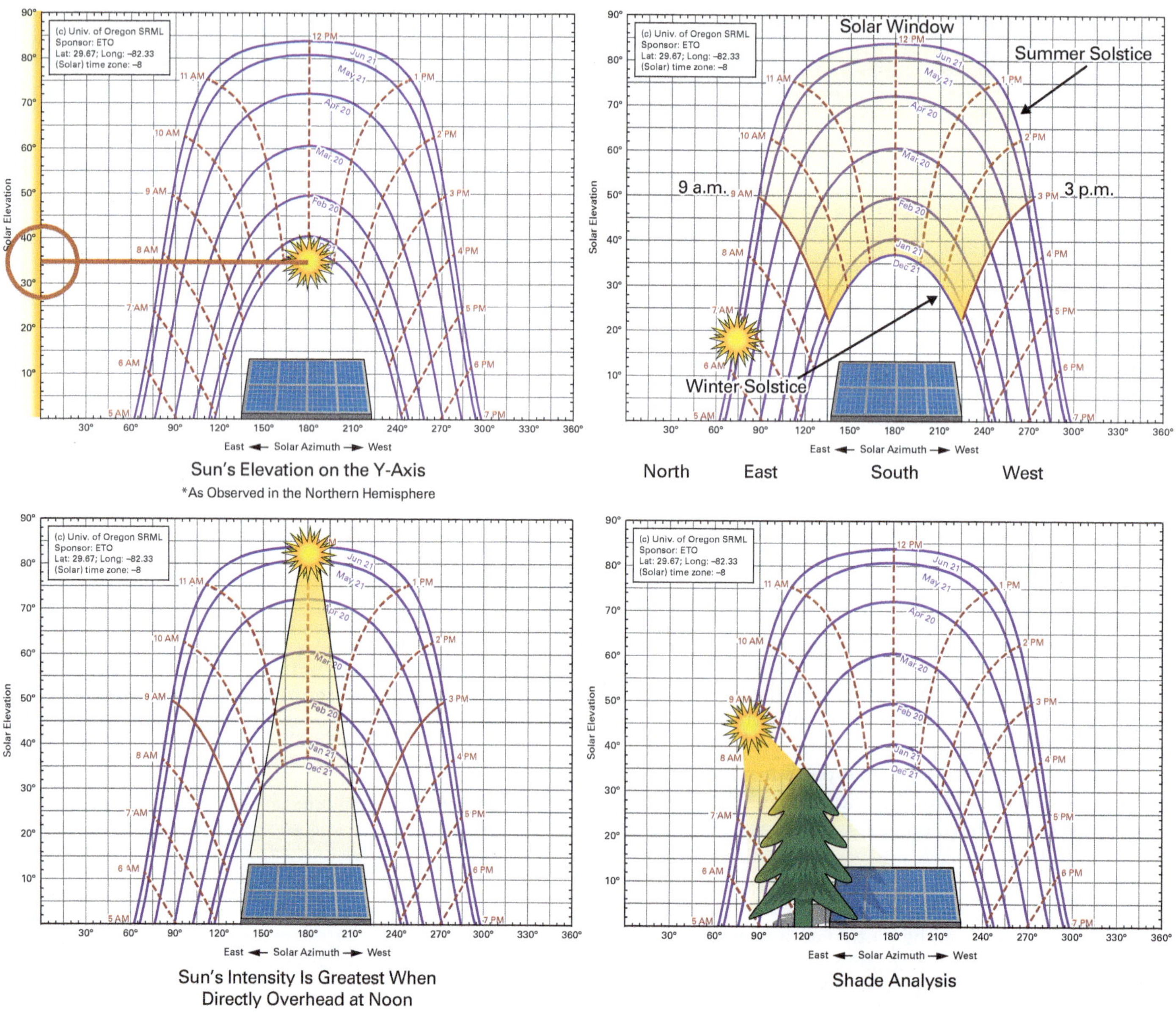

Figure 14 Sun path charts.

Source: University of Oregon, UO SRML: Sun chart program (uoregon.edu)

Recall that the optimal solar access is visualized as a solar window. The solar window is the Sun's annual movement between the hours of 9 a.m. and 3 p.m. and bounded by the Sun's highest and lowest points in the sky within the year, summer solstice, and winter solstice, respectively. Characteristics of the solar window include the following:

- The Sun's intensity is greatest when directly overhead.
- The Sun is above the tree line and less likely to cast shadows on the PV modules.
- The angle of the sunlight striking the PV module is more likely to be absorbed by the material, instead of reflected.

In the Northern Hemisphere, the optimal solar array position faces true south, and in the Southern Hemisphere, the optimal solar array position faces true north.

The observed sun path changes based on where you are located on the planet. *Figure 15* shows sun paths for Alachua, Florida, Little Falls, Minnesota, and Brasília, Brazil.

Compare the Sun's elevation, or *altitude*, in the summer at noon. Referring to both sun path charts from the Northern Hemisphere, note that the Sun's altitude is about 84° in Florida, while the Sun's altitude is about 66° in Minnesota during the summer solstice (June 21).

Sun Path Chart Applications

Several sun path chart applications are available. Choose your favorite and install it on your smartphone. Follow the application's instructions on entering your location's latitude and longitude to create a graph.

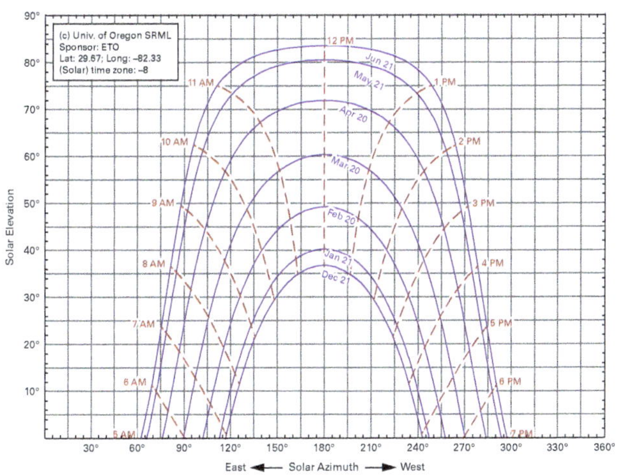

Alachua, Florida
(Northern Hemisphere)

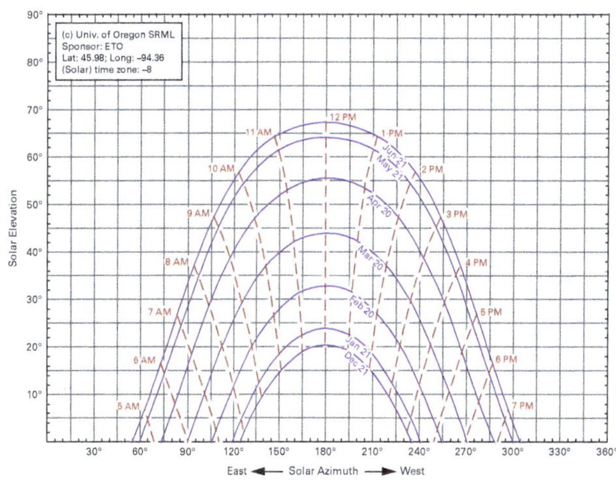

Little Falls, Minnesota

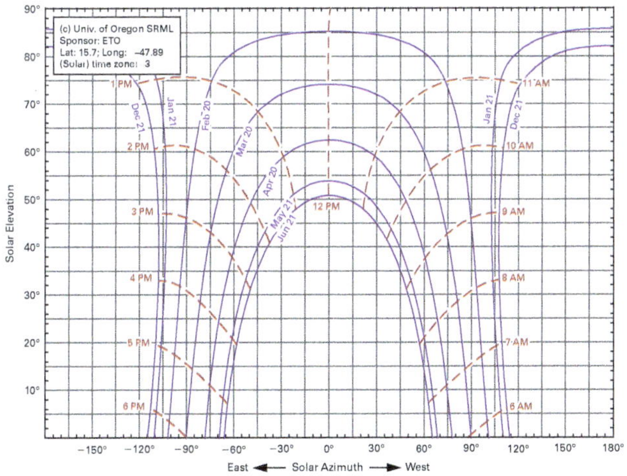

Brasília, Brazil
(Southern Hemisphere)

Figure 15 Sun path charts from different locations.
Source: University of Oregon, UO SRML: Sun chart program (uoregon.edu)

In general, optimize the position of the PV modules for the jobsite's location on Earth. The Sun's intensity is greatest when directly above an object and without obstructions such as clouds, dirt, or shade. Summer sunlight is much more intense than winter sunlight because it is closer to being directly overhead.

Positioning PV modules perpendicular to the Sun enables you to accomplish the following:

- Collect sunlight when the Sun's intensity is at its highest
- Absorb sunlight more efficiently than at other angles due to some limitations to the PV modules' materials

Software is used to simulate PV module performance throughout the year.

Sunlight Intensity

Peak sun hours: Solar irradiance a given location receives when the Sun's intensity is highest for a specific number of hours.

Sunlight intensity is measured in watts per square meter (W/m^2). One square meter equals 10.7639 square feet. Solar irradiance is the amount of solar radiation that strikes Earth's surface in a given span of time ($kW/m^2/days$). The term **peak sun hours** refers to the solar irradiance a given location receives when the Sun's intensity is highest for a specific number of hours ($kWh/m^2/day$). Peak sun hours, or *insolation values*, represent the equivalent number of hours per day when solar irradiance averages $1{,}000W/m^2$ for one hour.

To estimate the total PV module energy yield, multiply the PV module wattage by the peak sun hours per day for the geographical area. The answer equals the estimated watt-hours produced per day for that PV module. The PV module wattage hours are located on the manufacturer's nameplate, typically located on the back of the PV module.

As shown in *Figure 16*, regions of California, Arizona, New Mexico, and Texas receive over 5.75 $kWh/m^2/day$ of solar irradiance, which is equivalent to 5.75 peak sun hours. The following equations estimate the energy produced by a 400W PV module for an array in El Paso, Texas.

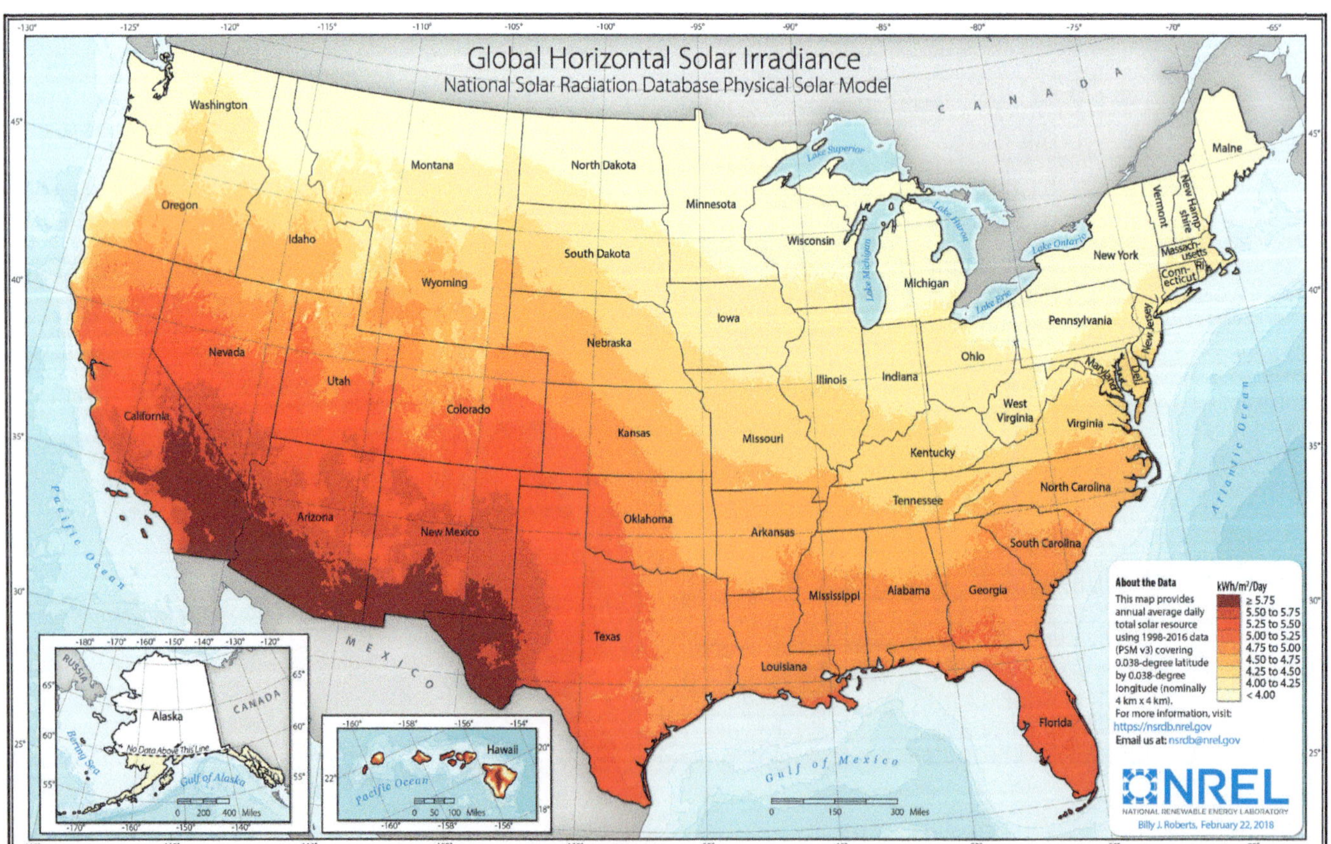

Figure 16 US solar irradiance.

Source: "The National Solar Radiation Data Base (NSRDB)," 2018. The National Renewable Energy Laboratory; U.S. Department of Energy.

$$\frac{5.75\ \text{kWh/m}^2\text{/day solar irradiance for El Paso, TX}}{1{,}000\text{W/m}^2\ \text{peak sun}} = 5.75\ \text{peak sun hours}$$

$$400\text{W PV module} \times 5.75\ \text{peak sun hours} = 2300\ \text{watt-hours (Wh) energy produced}$$

$$2300\ \text{watt-hours}/1000 = 2.3\text{kWh energy produced per day}$$

If the jobsite is located in the mountains, there are other things to consider. The higher the location, or elevation gain, the fewer gaseous molecules are in the air. In summer it has been observed that for every 1,000 feet of elevation gain, the Sun's ultraviolet (UV) rays intensify by 6% to 10% because the thinning atmosphere creates less solar obstruction. This effect results in the potential increase of solar irradiance to 1,250W/m^2 or greater per hour. Under clear sky conditions, a solar system at high elevation may outperform a solar system at sea level during certain times of the year, assuming the solar cells do not overheat.

1.4.3 Align with the Sun

PV modules are most effective when they are positioned as follows:

- Facing true south in the Northern Hemisphere, or facing true north in the Southern Hemisphere
- Tilted towards the Sun by adjusting the tilt angle (top-bottom) and azimuth angle (left-right)
- Free from obstructions that shade the modules

Most PV modules are mounted in a fixed-tilt position, but larger installations may invest in tracking systems with single- or dual-axis trackers that move the array throughout the day to follow the Sun.

NOTE

Single-axis and dual-axis trackers have special alignment considerations, which are covered in NCCER Module 57108, *Ground-Mounted Solar PV Installations*.

Face True South

In the Northern Hemisphere, based on observations of the Sun, positioning PV modules begins by pointing them towards true south to maximize the PV module exposure to the Sun. True south is the South Pole at the southern axis of the globe. True north is located at the North Pole, the northern axis of the globe. Both true south and true north are used in maps and calculations.

A compass (*Figure 17*) is used to find magnetic north. Magnetic north is located somewhere on Ellesmere Island in Northern Canada and is associated with Earth's magnetic field. It is a known location with an observable characteristic that true north and true south can be measured from.

Magnetic declination is the difference between magnetic north and true north and changes based on the observer's location on Earth.

To position the PV modules, use the following steps to find true south:

Step 1 Visit the National Oceanic and Atmospheric Administration (NOAA) website and find the magnetic declination for your location.

Step 2 Use a compass to find and mark magnetic north by aligning the needle with the N on the bezel.

Step 3 Find true north by rotating the compass base so that the needle is aligned with the magnetic declination value found on the compass' bezel.

Step 4 Find true south by adding 180° to true north.

Now that the general direction is established, a PV module can be tilted in two planes (*Figure 18*):

- Tilt angle (up-down)
- Azimuth angle (left-right)

Magnetic declination: The angle in degrees between true north and magnetic north that is specific for an observer's location on Earth.

Step 1

Look up the magnetic declination for your location with either your address or latitude and longitude.

Alachua, Florida
Latitude: 29.7938° N
Longitude: 82.4944° W
Magnetic Declination: 6°12' W

Step 2

Open a compass app. The red needle aligns with magnetic north.

Rotate to align the red needle with the N for north.

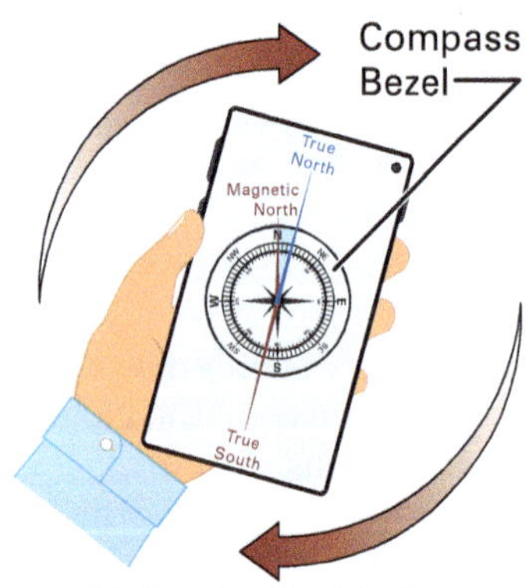

* Visualized with the orange line above.

Step 3

To find True North and True South, use the numbers on the compass bezel to rotate by the magnetic declination value.

If the value has a W, then rotate clockwise. If not, then rotate counterclockwise.

Rotated 6° 12' W
Magnetic North = 354°
True North = 0°
True South = 180°*

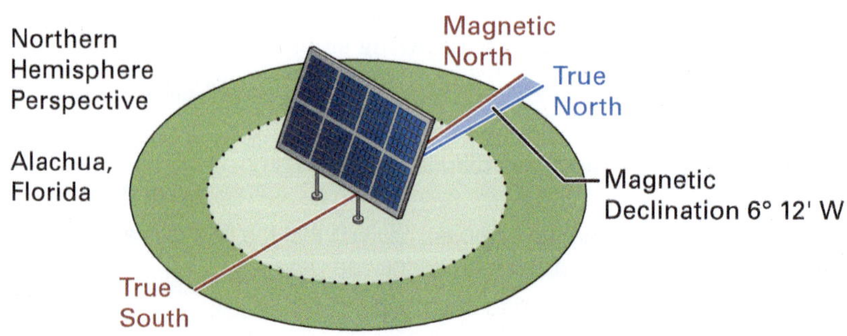

Magnetic North is roughly six degrees west of True North.

Figure 17 Magnetic north and true north for Alachua, FL.
Source: Data from https://www.ngdc.noaa.gov/geomag/calculators/magcalc.shtml

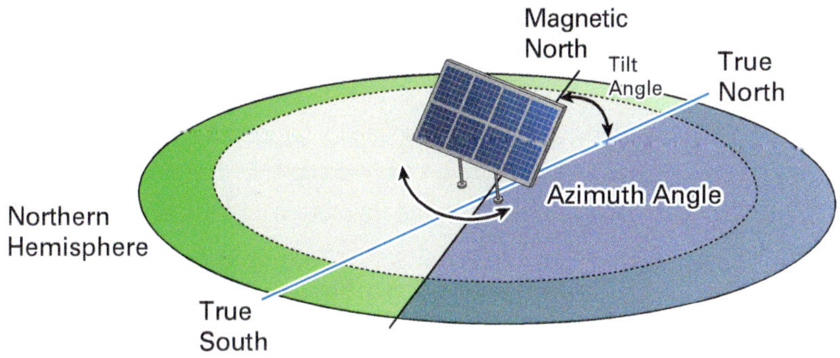

Figure 18 Solar PV module tilted in two planes.
Source: Based on https://www.ifc.org/wps/wcm/connect/ a1b3dbd3-983e-4ee3-a67b-cdc29ef900cb/ IFC+Solar+Report_Web+_08+05.pdf?MOD=AJPERES&CVID=kZePDPG

Tilt Angle (Up-Down)

Tilt angle is the pitch of the PV module when compared to a horizontal plane. It is expressed in degrees.

It is optimized along a chosen sun path within the sky, associated with time of year: summer, winter, or spring/fall.

A PV module's optimal position is for its surface to be perpendicular (90°) to the Sun (*Figure 19*). There are two methods to determine the PV module's tilt angle.

Tilt angle: In relation to solar PV, the up-down tilt position of a PV module or array.

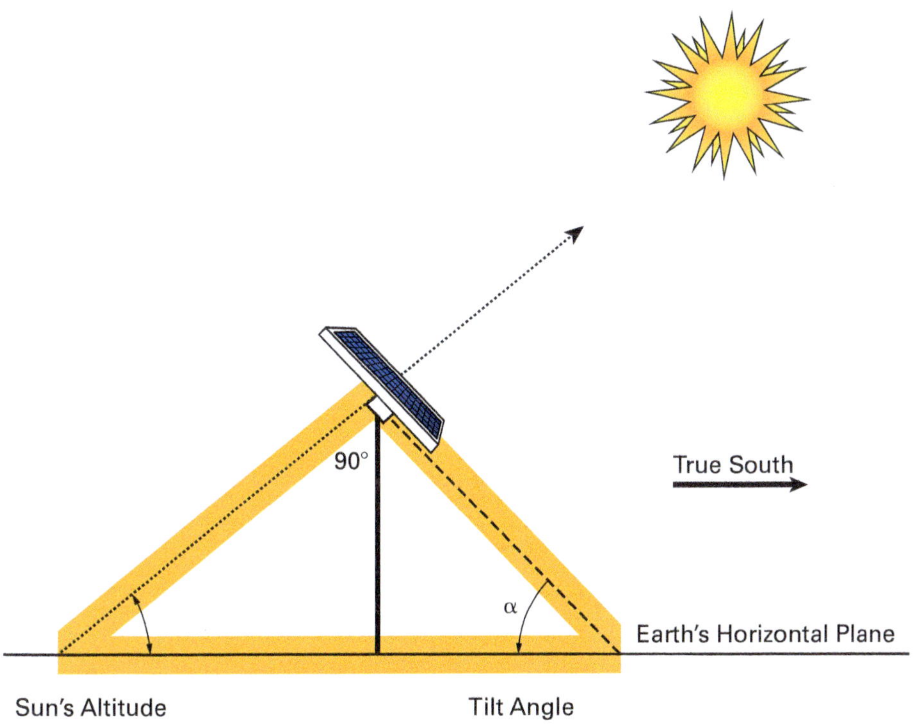

Figure 19 Adjust the tilt angle to align the solar PV module perpendicularly to the Sun.

How to Determine the Tilt Angle: Method 1

Use a sun path chart to choose a sun path and obtain the Sun's altitude. Because a PV module's optimal position to the Sun is 90°, the objects' orientation to one another creates a triangle; use the Sun's altitude to solve for the tilt angle.

Choose a Sun Path

The customer's needs will provide reasons to optimize to a winter, summer, or spring/fall sun path.

Step 1 Using a sun path chart for the jobsite location, select a sun path.

Step 2 Obtain the Sun's altitude, located on the Y-axis of the chart.

Step 3 Solve the right-triangle equation for the PV module's tilt angle.

Using the sun path chart for Little Falls, Minnesota in *Figure 15*, let's look at an example. After weighing the options, the homeowner(s) chose to optimize their home system to maximize energy production based on the available sunlight on the shortest day of the year, the winter solstice. It is a time when there are fewer hours of sunlight, longer shadows, and the potential of bad weather. Their energy goal is to produce the most energy on the anticipated worst day of the year. Looking at the sun path chart, find the Sun's elevation, or *altitude,* that corresponds with Dec 21. This should be approximately 20°.

As shown in the following equation, use the Sun's altitude to solve for the tilt angle:

$$\text{Tilt angle} = 90° - \text{the Sun's altitude}$$

How to Determine the Tilt Angle: Method 2

The spring/fall sun path is considered the happy medium to provide maximum output power on an annual average. Earth's position with the Sun makes figuring out the PV module's tilt angle easy.

Declination: The angle between Earth's Equator and a celestial object, like the Sun.

Latitude: A measurement expressed in degrees for a location's distance in relation to the Equator.

The **declination** of the Sun is the angle between the Equator and the Sun. During the Vernal (spring) Equinox (on or about March 21st) and the Autumnal (fall) Equinox (on or about September 21st), the angle is 0°, positioning the Sun directly above the Equator.

Latitude is a measurement expressed in degrees for a location's distance to the Equator. For example, NCCER is located at latitude 29.8° N (north of the Equator). Brasília, Brazil is located at 15.8° S.

Conveniently, this makes one's latitudinal location on the planet (*Figure 20*) an optimum tilt angle for the PV module, as the Sun's rays strike the Equator in March and September.

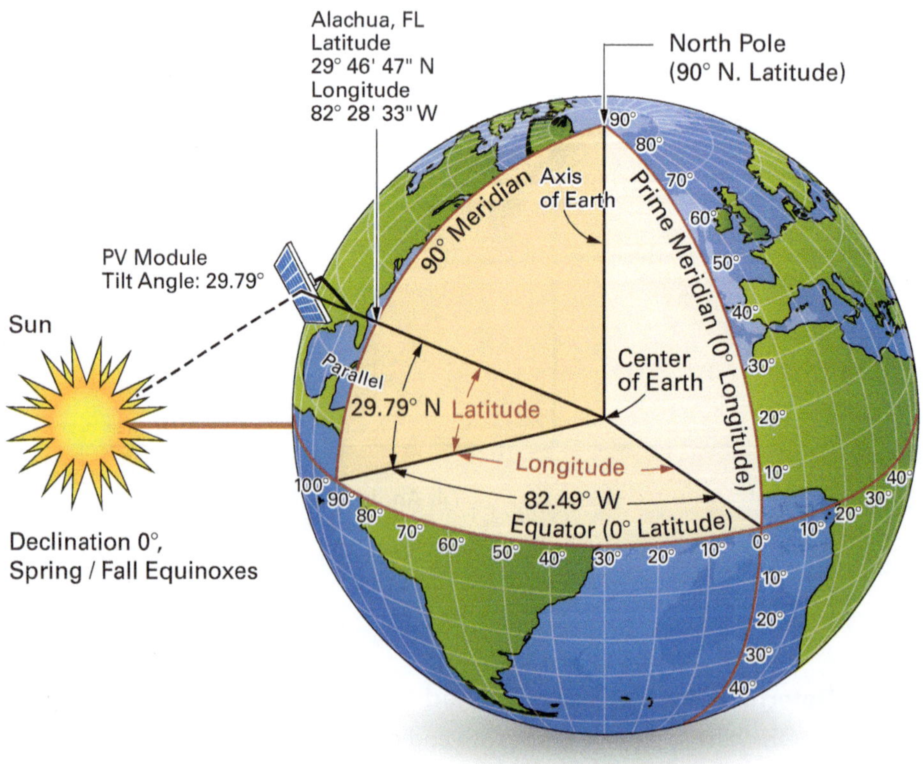

Figure 20 Relationship of declination and latitude.

With this method, it becomes unnecessary to solve the triangular equation or use the sun path chart for obtaining the Sun's altitude. Instead, look up your location's latitude with an online tool; for example, a Google search for "little falls, mn latitude."

Because declination changes throughout the year, the optimal PV module tilt angle also changes. If you know your location's latitude, then that value can be used to seasonally adjust the tilt angle (*Figure 21*). (Note that because the latitude is the known value, the representation of the tilt angle has shifted to a different corner within our triangle.) Optimal tilt angles for each season are as follows:

- Winter tilt angle = jobsite's latitude +15°
- Summer tilt angle = jobsite's latitude −15°
- Spring/fall tilt angle = jobsite's latitude

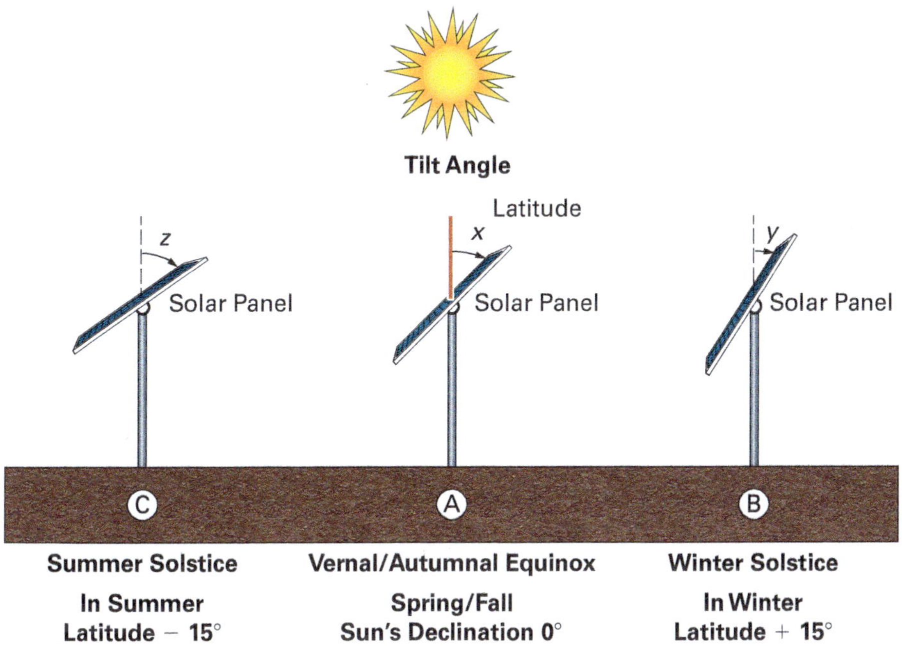

Figure 21 Seasonally adjust the tilt angle +/−15° for winter or summer.

Azimuth Angle (Left-Right)

The PV module's left-right positioning is called the **azimuth angle** (*Figure 22*), named for its alignment with the Sun's azimuth.

The azimuth is defined by the *Merriam-Webster Dictionary* as "an arc of the horizon measured between a fixed point (such as true north) and the vertical circle passing through the center of an object, usually in astronomy and navigation clockwise from the north point through 360 degrees."

Azimuth angle: In relation to solar PV position, the angle clockwise from true north for setting the left-right rotation of a PV module.

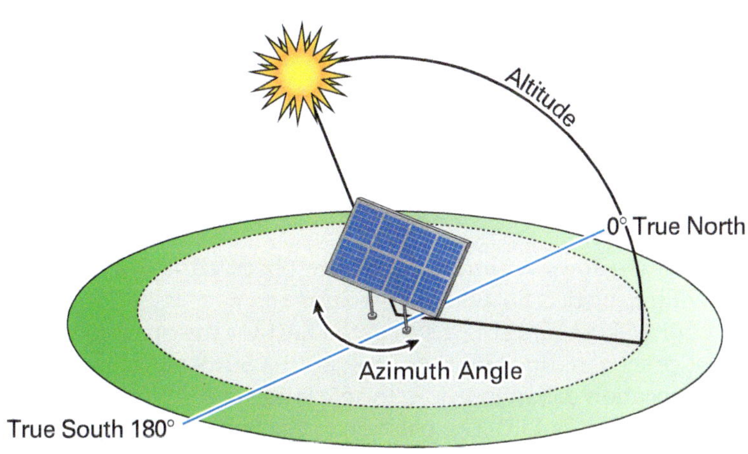

Figure 22 Azimuth angle of a PV module.

The azimuth angle optimizes the PV module position with the Sun for the time of day. In the Northern Hemisphere, setting azimuth to true south (180°) aligns with the solar window. This alignment is required for solar rebates and financial analysis for customer loan approvals.

1.4.4 Shade

To model the effects of shading on a jobsite, modeling software is recommended to generate a 3D representation of the landscaping and shading.

Modeling solar design software, such as HelioScope and Aurora by Aurora Solar® and Terabase, overlays solar irradiance (*Figure 23*) and shade analysis data for a given US address to visualize the optimal PV module positions for generating electricity.

Figure 23 Building overlay of solar irradiance data for optimization of solar PV module placement.
Source: Aurora Solar

Recall that the Sun is lowest in the sky—casting the longest shadows—in the winter. Consider making shade estimates using the Sun's position during the winter solstice.

During a site assessment, document the height of objects, such as roofing ventilation systems and parapets. Solar system design software uses these details to reduce the number of suboptimal PV modules in the design.

Shade analysis can also be performed manually at the jobsite. The height of trees or any object that may shade the proposed installation site is measured using a clinometer, hypsometer, or Abney level and documented on the sun path chart (*Figure 14*).

Inter-Row Spacing

Ground-mounted solar installations require inter-row spacing considerations. As shown in *Figure 24*, the rows of PV modules are positioned so the shade from one row doesn't overlap the next.

The longest shadows within the year are observed at 9 a.m. and at 3 p.m. along the winter solstice. So, where's the Sun?

As shown in *Figure 25*, using a sun path chart for the given location, find the winter solstice's 9 a.m. line and then look up the Sun's elevation at that time of day. For this location in San Jose, CA, the elevation is 25°. Next, look up the Sun's azimuth angle, and it is 135°. The position of the Sun will cast the shadow at an angle to the PV module, rather than hitting the module's face square on.

At this morning hour, the Sun casts its shadow on the ground (Earth's horizontal plane), but it is offset from the rows by the azimuth angle. There is enough

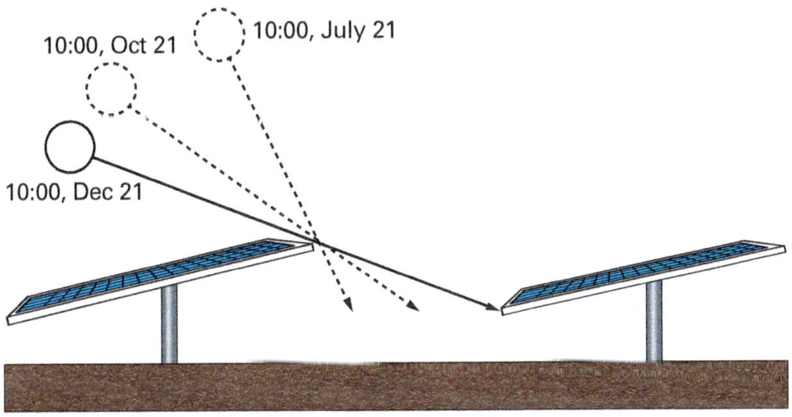

Figure 24 Inter-row shading influenced by the position of the Sun.

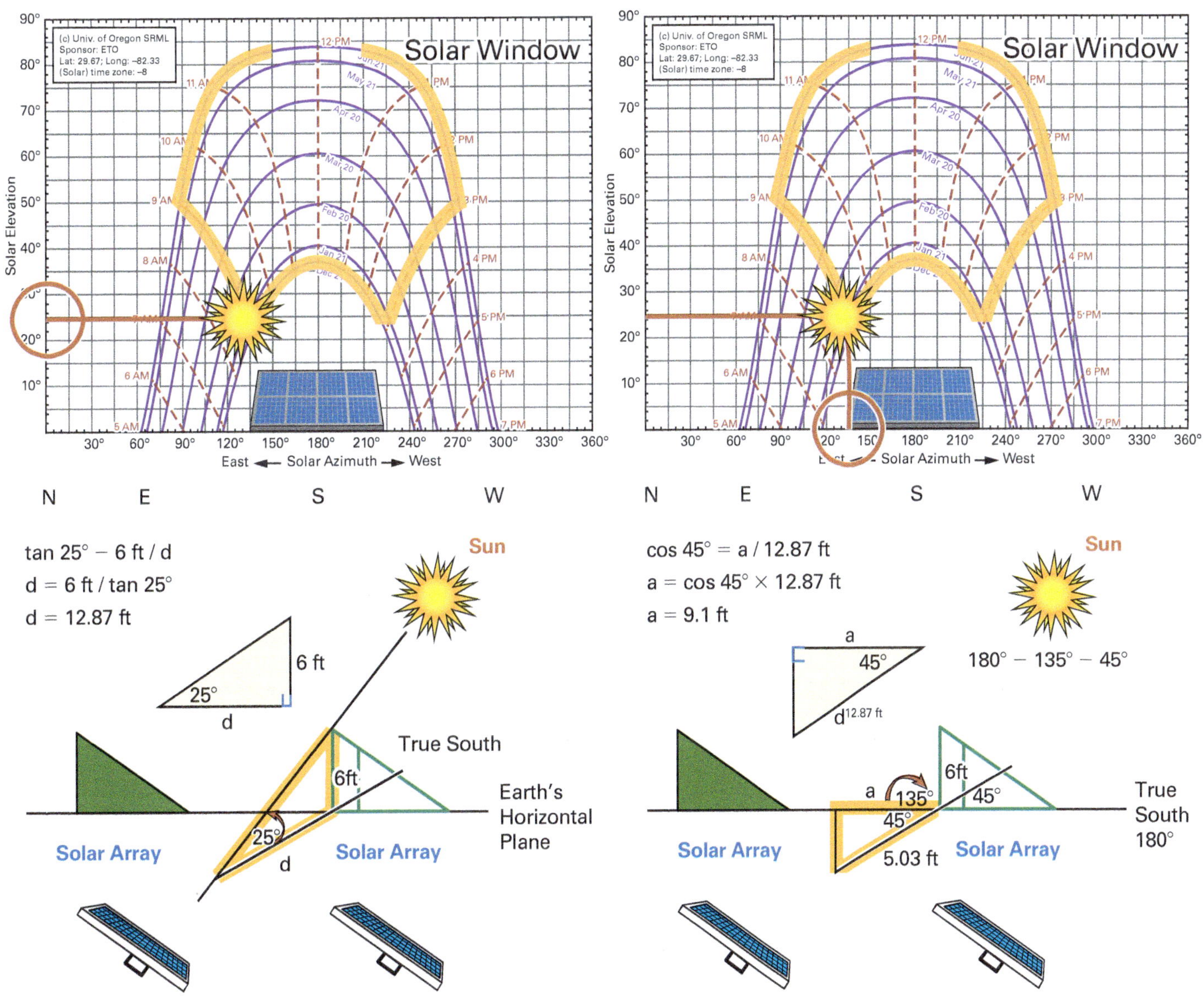

Figure 25 Calculating inter-row spacing.

information to solve for the length of the bottom edge of that triangle. Using a trigonometry equation, the triangle's side along the ground (d) can be solved. Side d is equal to 6 feet divided by the tangent of 25°, the Sun's elevation. Side d is 12.87 feet.

What is really needed is the length of the shadow between the two rows (a), but we have (d) from the first equation. Now, shifting planes to the ground, the leg of the triangle becomes the length of the hypotenuse of the second triangle.

Using a second trigonometry equation, the cosine of the 45° angle is equal to the adjacent (a) over the hypotenuse (d). Solving for side a, the cosine of 45° multiplied by 12.87 feet is equal to 9.1 feet. Based on the calculations, position the second row 9.1 feet from the first to avoid shading.

1.5.0 Electrical Theory

Ampacity: The maximum current in amperes a conductor can carry continuously under the conditions of use without exceeding its temperature rating.

Electricity is a form of energy that can be used by electrical devices, such as motors, lights, TVs, heaters, and numerous other devices, to perform work.

An electrical circuit contains, at minimum, a voltage source, a load, and conductors (wires) to carry the electrical current. The circuit should also have a means to stop and start the current, such as a switch.

A common analogy (*Figure 26*) used for visualizing the flow of electricity, or current, through a circuit is the flow of water through a hose.

The electrical current is measured in amperes, or amps. It is the amount of electrons that flow past a given point each second.

Voltage, measured in volts, pushes the electrons through the circuit, as pressure.

A conductor, like wire, is the hose. The larger the opening in the hose or larger gauge of wire, the less resistance on the given current. It is easier for the electricity to flow.

Ampacity is the maximum current in amperes a conductor can carry continuously under the conditions of use without exceeding its temperature rating.

When the ampacity remains the same, but the resistance increases or the gauge of wire is decreased, then the wire begins to heat up and may melt. For this reason, selecting a wire gauge for the amps in the circuit is critical. Resistance is measured in ohms (Ω).

The amount of energy a load uses is called power, and it is expressed in watts. Volts (V), amperes (A), watts (W), and ohms (Ω) are related in such a way that if any one of them changes, the others are proportionally affected.

1.5.1 Watt's Law

Watt's law is an equation that explains the relationship of power, current, and voltage. It can be used to help select the proper wire gauge for a solar PV installation, among other things.

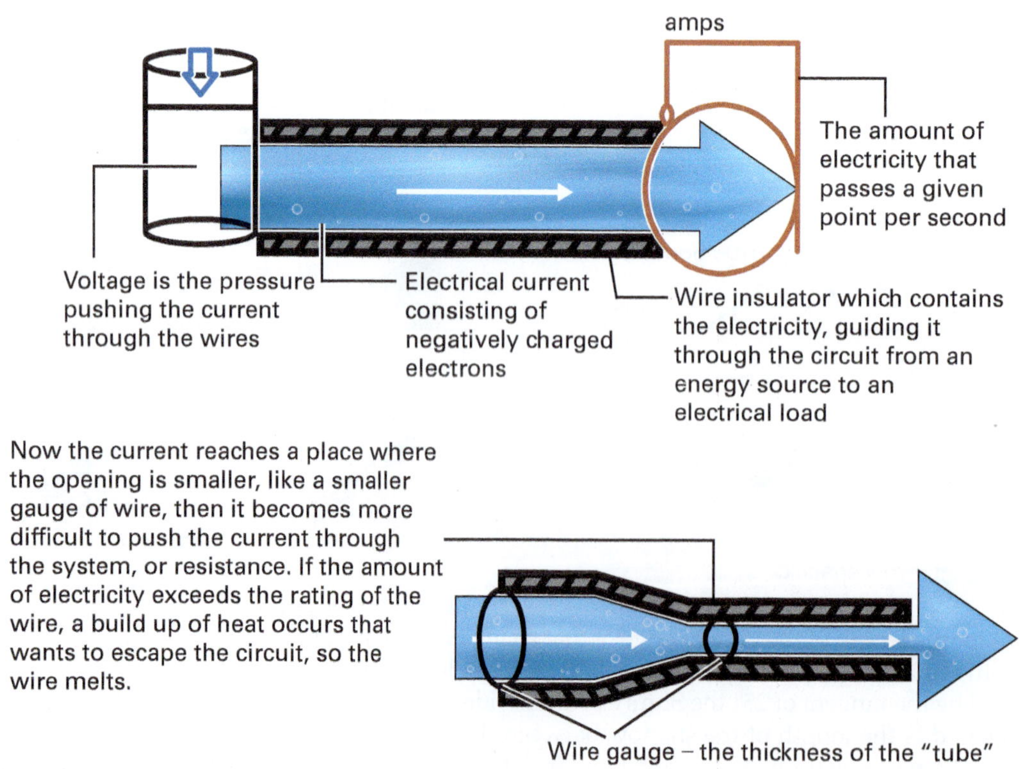

Figure 26 Water analogy.

Power is the ability of an electrical load, like a light bulb, to do work. In electrical circuits, power is measured in watts (W). One watt is the power used when one ampere of current flows through a potential difference of one volt.

As shown in the following equation, Watt's law states power (P) is determined by multiplying the rated current (I) by the rated voltage, or *electromotive force* (E):

$$\text{Power(P)} = \text{Current(I)} \times \text{Voltage(E)}$$

Using variations of this equation, the power, voltage, or current in a circuit can be calculated whenever any two of the values are known.

To help remember this equation, you can see the relationships of the elements in *Figure 27*. If you want to solve for power, place your thumb over the power in watts and you see $P = I \times E$. If you want to solve for current in amps, then you place your thumb over the I and you see $I = P/E$.

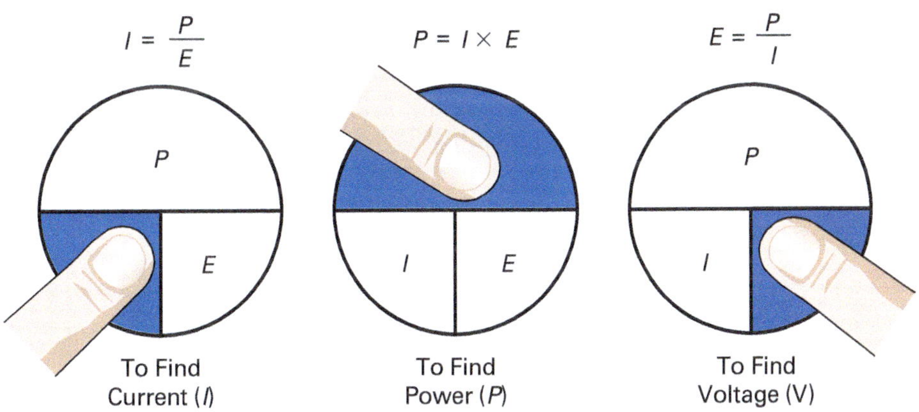

	Letter Symbol	Unit of Measurement
Current	*I*	Amperes (A)
Power	*P*	Watts (W)
Voltage	*E*	Volts (V)

Figure 27 Watt's law.

Watt-hours (Wh) are calculated by multiplying the power in watts (W) by the number of hours during which the power is used. The kilowatt-hour (kWh) is commonly used for larger amounts of electrical work or energy. The prefix *kilo* means one thousand. For example, if a light bulb uses 100W or 0.1kW for 10 hours, the amount of energy consumed is 0.1kW × 10 hours = 1.0kWh.

Watt-hours (Wh): A unit of energy that describes the amount of power used over time, typically used for metering.

Very large amounts of electrical work or energy are measured in megawatts (MW). The prefix *mega* means one million.

Information to help you with calculations can be found on the PV module's nameplate, located on the PV module's **back sheet**. This is the back of the PV module.

Back sheet: The back of a solar photovoltaic (PV) module.

As an example, suppose the total load current in the main line equals 20A and the voltage is 120V. Using Watt's law, how much power is generated?

$$P = I \times E$$
$$P = 20\text{A} \times 120\text{V}$$
$$P = 2400\text{W } \textit{or } 2.4\text{kW}$$

If this power is used for five hours, then how much energy is used over time?

$$2.4\text{kW} \times 5 \text{ hours} = 12\text{kWh}$$

1.5.2 Ohm's Law

Ohm's law defines the relationship between current (I), voltage (E), and resistance (R). If any two of the values are known, the third can be calculated. Resistance is a force within the circuit which reduces the electricity's ability to flow in the circuit.

Ohm's law is an equation used when you are figuring out the voltage drop within a circuit, for example.

There are three ways to express Ohm's law mathematically:

- The current (I) in a circuit is equal to the applied voltage (E) divided by the resistance (R).
- The resistance (R) of a circuit is equal to the applied voltage (E) divided by the current (I).
- The voltage (E) applied to a circuit is equal to the product of the current (I) and the resistance (R).

The Ohm's law equations can be practiced using an Ohm's law circle, as shown in *Figure 28*. To find the equation for E, I, or R when two quantities are known, cover the unknown third quantity. The other two quantities in the circle indicate how the covered quantity may be found.

A few Ohm's law examples are shown in *Figure 29*.

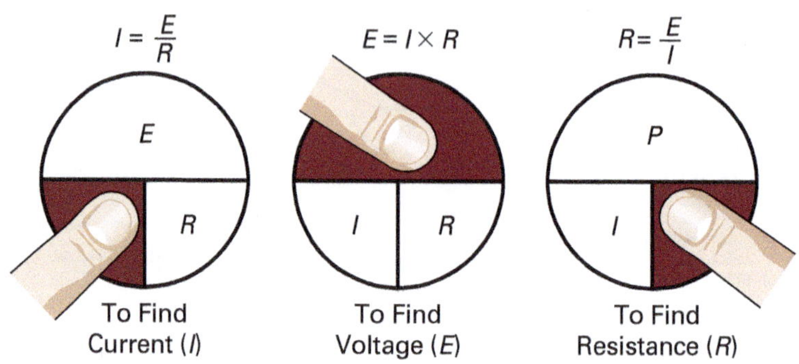

	Letter Symbol	Unit of Measurement
Current	*I*	Amperes (A)
Resistance	*R*	Ohms (Ω)
Voltage	*E*	Volts (V)

Figure 28 Ohm's law circle.

Find *I* when *E* = 120V and *R* = 30 ohms.

Find *R* when *E* = 240V and *I* = 20A.

Find *E* when *I* = 15A and *R* = 8 ohms.

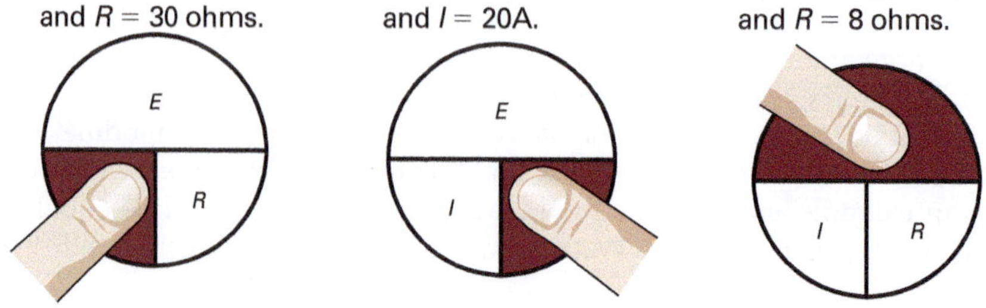

$I = \frac{E}{R}$

$I = \frac{120V}{30\Omega}$

$I = 4A$

$R = \frac{E}{I}$

$R = \frac{240V}{20A}$

$R = 12\Omega$

$E = I \times R$

$E = 15A \times 8\Omega$

$E = 120V$

Figure 29 Ohm's law examples.

1.5.3 Circuits

PV modules, batteries, and electrical loads are added to circuits by wiring them in series, in parallel, or in series-parallel.

WARNING!

If the amperage produced by the PV modules exceeds the inverter's limits and/or wire ampacity, then the wires may melt and cause a fire. Know the limits to the number of PV modules that can be strung together.

Watt's law can be used to determine the amount of power supplied by the solar PV system by multiplying the system's voltage by its amperage. Wiring strategies can increase the wattage of the system when adding power sources, such as PV modules and batteries, to reach a desired electrical output. Wiring in series increases the voltage in the summation of each PV module in the circuit, while the amperage remains constant across the circuit. Wiring in parallel increases the amperage in the summation of each PV module, while the voltage remains constant. *Figure 30* shows the following two ways to visualize the information:

- Visualize the PV module wiring when PV modules are added to the mounting hardware.
- Visualize the math.

Series

System components wired in series have one path that the electricity travels along. When PV modules are wired in series, the positive terminal of one PV module is connected to the negative terminal of another.

In the example shown in *Figure 30*, a residential solar kit was chosen from Amazon. The PV module's maximum power at **Standard Test Conditions (STC)** is 320W of power. This information is found on the PV module's nameplate. Watt's law confirms this by multiplying the optimum operating voltage (32.7V) by the optimum operating current (9.79A).

$$32.7\text{V} \times 9.79\text{A} = 320\text{W}$$

When determining wiring strategies, always design to the maximum limits of the PV modules. Use the PV module's open circuit voltage and the short circuit current when calculating PV string output. In the example, two 40.1V/10.36A PV modules at STC wired in series will produce 80.2V at 10.36A or 830.87W at STC. Four 40.1V/10.36A PV modules wired in series will produce 160.4V at 10.36A or 1661.74W at STC.

However, these PV modules are not feasible in accordance with *NEC®* residential safety guidelines which require the open circuit voltage to be less than 40V. The message here is to know the code before selecting product for the intended application.

Parallel

System components wired in parallel have multiple paths that the electricity can travel along. When PV modules are wired in parallel, the positive terminal of one PV module is connected to the positive terminal of the next PV module. As they generate electricity, for every module in the parallel circuit the amperage is additive, but the voltage stays the same. For example, two 40.1V/10.36A PV modules wired in parallel will produce 40.1V at 20.72A. Four 40.1V/10.36A PV modules wired in parallel will produce 40.1V at 41.44A.

Series-Parallel

Solar arrays and batteries can be connected in series-parallel to achieve the desired, or maximum output, voltage and current of the total array (*Figure 30*). For example, two 40.1V/10.36A PV modules can be wired in series to produce 80.2V at 10.36A, and then connected to a second group of PV modules wired in

Standard Test Conditions (STC): Controlled laboratory environment specifications for PV module operating temperature, solar irradiance, and air mass used in standardizing PV module performance ratings.

NOTE

STC represents the laboratory environment for testing PV module performance. The environment is controlled to specific operating temperature, solar irradiance, and air mass so that the performance ratings of one PV module product can be compared to another from a different manufacturer for product selection.

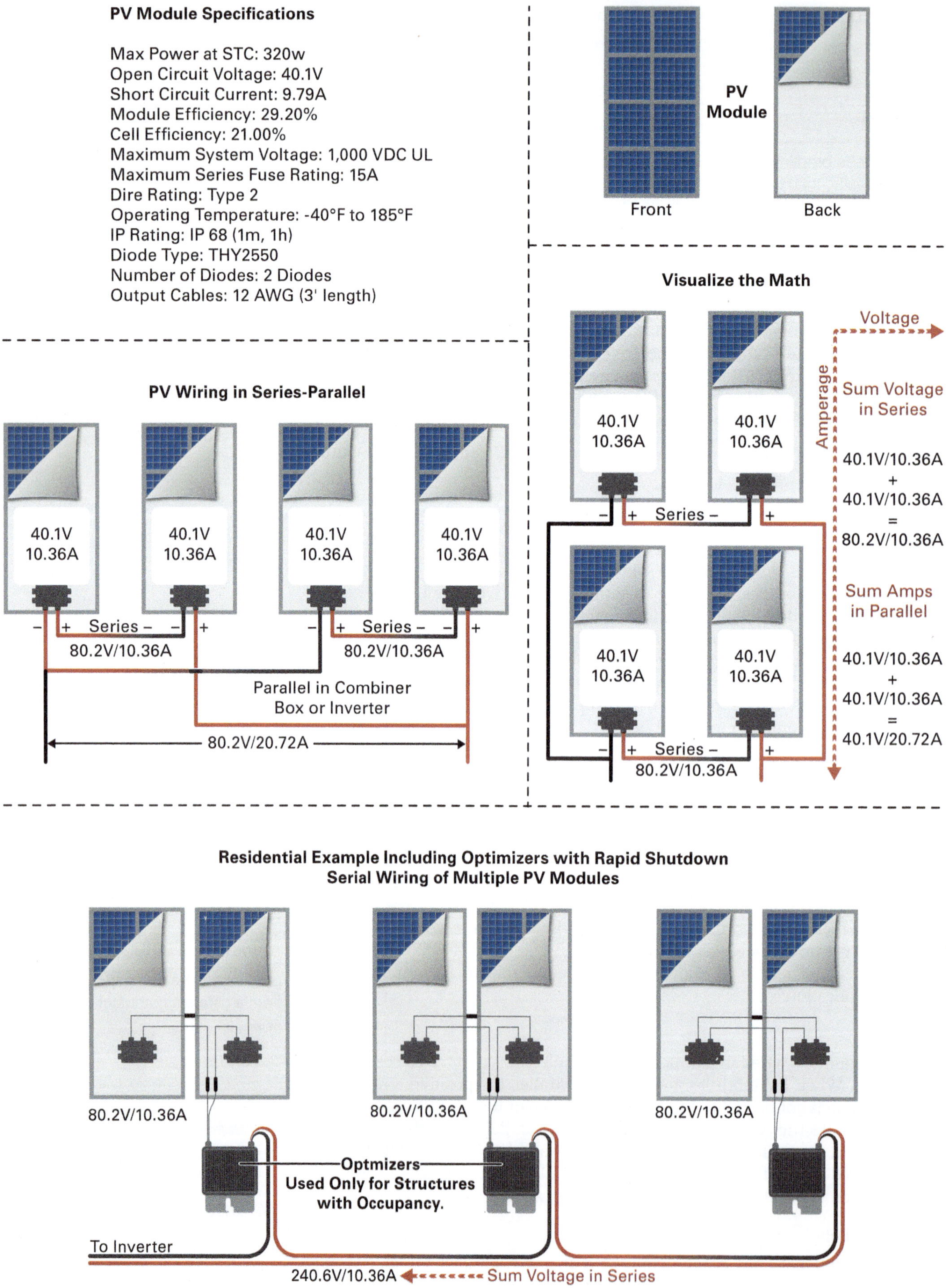

Figure 30 Wiring solar PV modules.

parallel to produce a total output of 80.2V at 20.72A. If these four PV modules are then wired in parallel to four other PV modules wired in the same way, the voltage will be 80.2V at 41.44A, and so on. This is the basic concept behind building solar arrays.

1.0.0 Section Review Questions

1. Which part of the PV module transforms solar energy into electrical energy by releasing negatively charged electrons?
   a. The battery
   b. The DC energy load
   c. The electrical connectors
   d. The semiconductor

2. Which *NEC*® article covers large-scale PV facilities?
   a. *Article 330*
   b. *Article 430*
   c. *Article 520*
   d. *Article 691*

3. Which of the following represents a typical residential system size?
   a. 1–3kW
   b. 3–10kW
   c. 10kW–2MW
   d. >2MW

4. Of these system designs, which one may include switchgear?
   a. Off-grid
   b. Grid-integrated
   c. Grid-tied
   d. Utility scale

5. In the Southern Hemisphere, while assessing a jobsite for optimal solar access, use a compass to establish basic array orientation with the Sun to point the PV modules towards _____.
   a. east
   b. magnetic north
   c. true north
   d. true south

6. What is the Sun's angle of declination during the Vernal (spring) Equinox?
   a. Zero
   b. 15 degrees
   c. 23.45 degrees
   d. 90 degrees

7. Electrical current is measured in which of the following units?
   a. Terminal
   b. Watts
   c. Voltage
   d. Amperes

2.0.0 Solar Photovoltaic (PV) Primary Components

Performance Tasks

This is a knowledge-based module. There are no Performance Tasks.

Objective

Describe PV system primary components.

a. Describe different types of mechanical components for ground-mounted and roof-mounted PV systems, including motion capabilities.
b. Explain general similarities and differences among the three main types of solar cells.
c. Explain how PV modules from various manufacturers can be compared for performance.
d. Describe the purpose of an inverter.
e. Understand the differences in battery options and requirements for an energy storage solution.

Solar arrays and systems consist of primary components and balance of system (BOS) components (*Table 4*), which generate, distribute, and control the flow of energy. Recall that not every system configuration requires all these components.

TABLE 4 Solar PV Components

Primary Components	Balance of System (BOS) Components

• PV module(s)
• Optimizers
• Inverter(s) or microinverter(s)
• Energy storage
  ○ Batteries
  ○ Charge controller
• Electrical load(s)
• Structural components (i.e., tracker systems, rail or racking systems, ground mounts, carport structures)
	• Conductors and wiring methods
	• Raceways and conduits
	• Junction and combiner boxes
	• Disconnect switches
	• Fuses and circuit breakers
	• Terminals and connectors
	• Grounding equipment
	• Array mounting and other structural hardware

The primary components are those that are selected and presented to an engineer to include in the solar design. These are typically expensive components that have manufacturing marketing campaigns to sell product. The engineer develops the plans and defines a list of parts, or balance of system components, necessary to execute the installation. The cost of BOS components can also be expensive. For example, the amount of wire that needs to be purchased and the excavation costs for running it underground quickly add up.

In Section One, you saw one-line diagrams of various system configurations. *Figure 31* helps you visualize the exterior and interior placement options of the PV components in relation to a building, to begin to understand the following spatial relationships:

- The size of the mounting surface to the size of the solar PV array
- The distances between the main components of the solar PV system
- The distance from the building to the connection with the power grid

In this section, you'll learn about the primary components that make up a solar PV system or array.

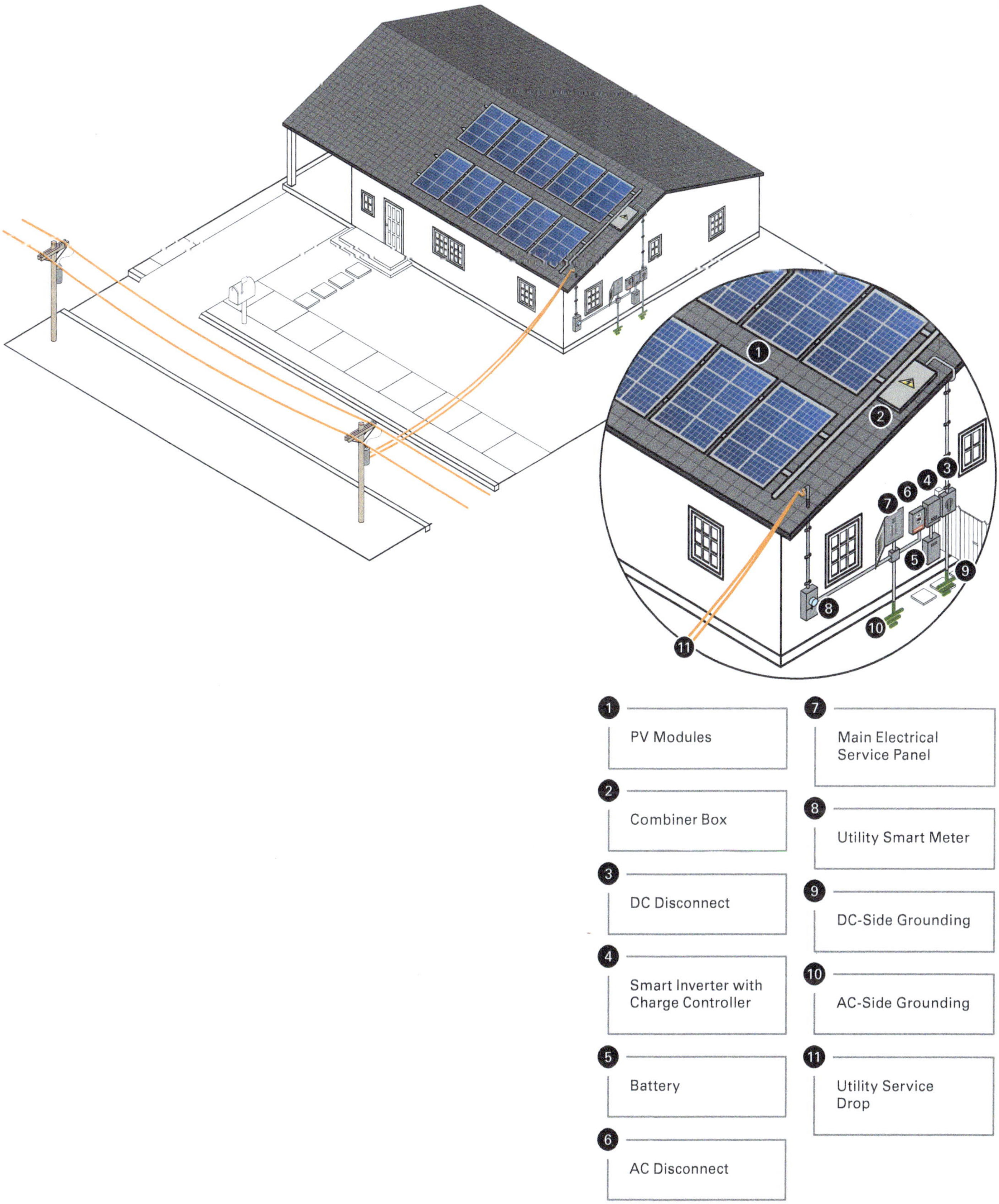

Figure 31 Visualize spatial relationships.

2.1.0 Mechanical System Components

The mechanical system components are the footers and support structures of the system. There are several solar PV mounting solutions that relate to the type of surface the solar system is mounted to (*Table 5*).

TABLE 5 PV Module Installation Options

Examples of Array Locations	Mounting Methods	Motion Capabilities
Ground: clay, dirt, rock, sand, soil	Pole mounted	Fixed tilt
Structure: concrete, roof decking, metal, wood	Ground-mounted	Tracker mounted
	Roof-mounted	Floating
Water: piers, piles	Integrally mounted	

* Not an exhaustive list

Building-integrated photovoltaics (BIPV): A PV system built into the structure as a replacement for a building component such as roofing.

When a PV module is built into a structure, it is known as **building-integrated photovoltaics (BIPV)**.

2.1.1 Motion Capabilities

Solar PV arrays can be fixed tilt or tracking mounted (*Figure 32*). Fixed tilt has one stationary alignment of the PV array with the Sun, which may be manually adjusted with the seasons. The tracking mount increases installation costs, but increases the energy yield of the solar system, leading to higher investment returns over time. It is seen most often at the utility scale.

Single-axis trackers move entire rows of PV modules to follow the Sun. The rows are installed in a north-south direction, so the tilt moves from east to west. They are the most common at utility scale.

Dual-axis trackers allow for modules to remain pointed directly at the Sun throughout the day. The dual movement increases energy yield, but the investment in the equipment is high.

(A) Fixed Tilt **(B) Single-Axis Tracker** **(C) Dual-Axis Tracker**

Figure 32 Motion.
Source: (A) Jerzy Niezabitowski/Getty Images; (B) Kate Scott/Shutterstock; (C) pedrosala/Getty Images

2.1.2 Ground-Mounted Systems

Ground-mounted installations require foundational work. Soil studies provide information, such as whether the soil is too corrosive or too hard and impenetrable. This information guides design decisions; for example, choosing to pour concrete footers versus using specially coated metal posts, or a ballasted solution with concrete blocks.

Once the foundation is established, the installers assemble the support structure, or *racking*, for the PV modules. This portion of the assembly is similar to rack-mount roof installations. Common parts are displayed in *Figure 33*.

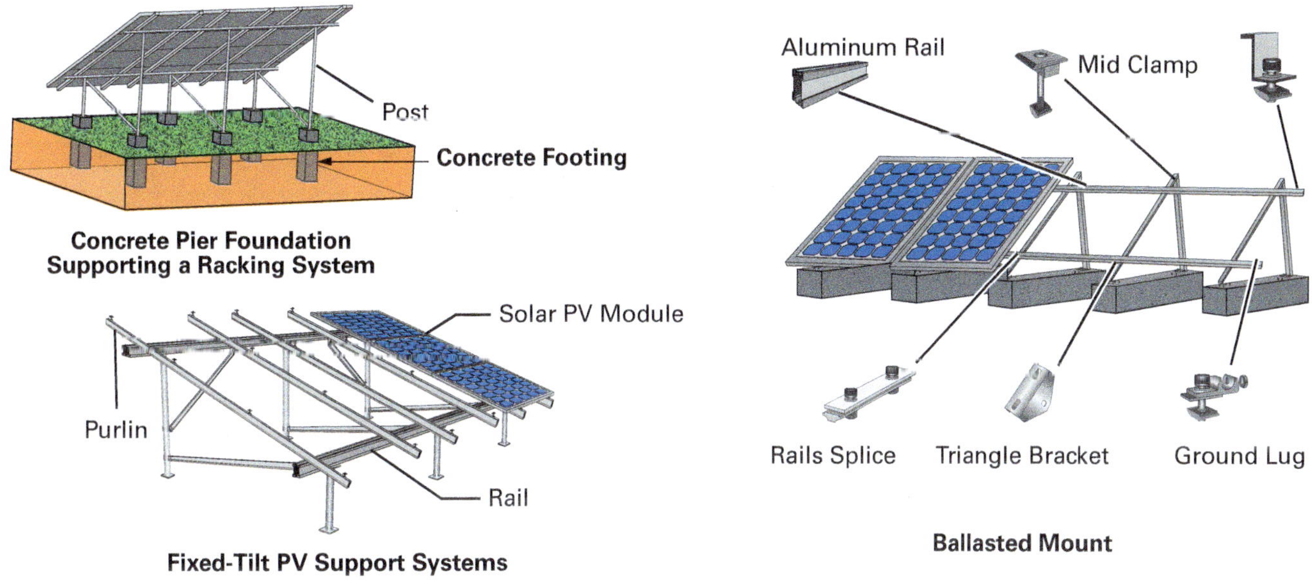

Figure 33 Fixed-tilt, common rack-mount parts.
Source: https://sunmodo.com/flat-roof-sunbeam/

Some example applications are solar carports and shade structures as shown in *Figure 34*. Benefits of these types of system designs include the following:

- Optimized use of space
- Protection of vehicles from the damaging effects of sunlight and weather
- Reduced PV module vandalism due to the height of the installation

CAUTION

Ground-mounted systems require security fencing, which is an electrical safety precaution and a deterrent from vandalism.

EV Charging Stations

For safety reasons, PV modules seen at an EV charging station do not assist in charging the car's battery or the recharge EV station batteries. The electricity goes directly to the power grid.

Figure 34 Ground-mounted installation at an EV charging station.
Source: Lorado/Getty Images

2.1.3 Roof-Mounted Systems

Solar PV arrays may be roof mounted. There are low-slope and steep-slope roofs. The NCCER Roofing Program defines a steep-slope roof as having a rise-to-run ratio of 4:12 or greater, or roof slope with an angle of at least 18.43° (*Figure 35*).

The roof slope and the roofing material influence how the PV modules are mounted (*Figure 36*).

For mounting on steep-slope roofs, PV module tilt angle typically follows the roof's slope. Additional tilt optimization by adding more hardware to the manufacturer's out-of-the-box product is usually not considered cost-effective.

NOTE

A solar PV roof installation requires an understanding of roofing fundamentals so you do not compromise the integrity of the entire roof system. More details are provided in NCCER Module 57106, *Roofing Fundamentals* and NCCER Module 57107, *Roof-Mounted Solar PV Installation*.

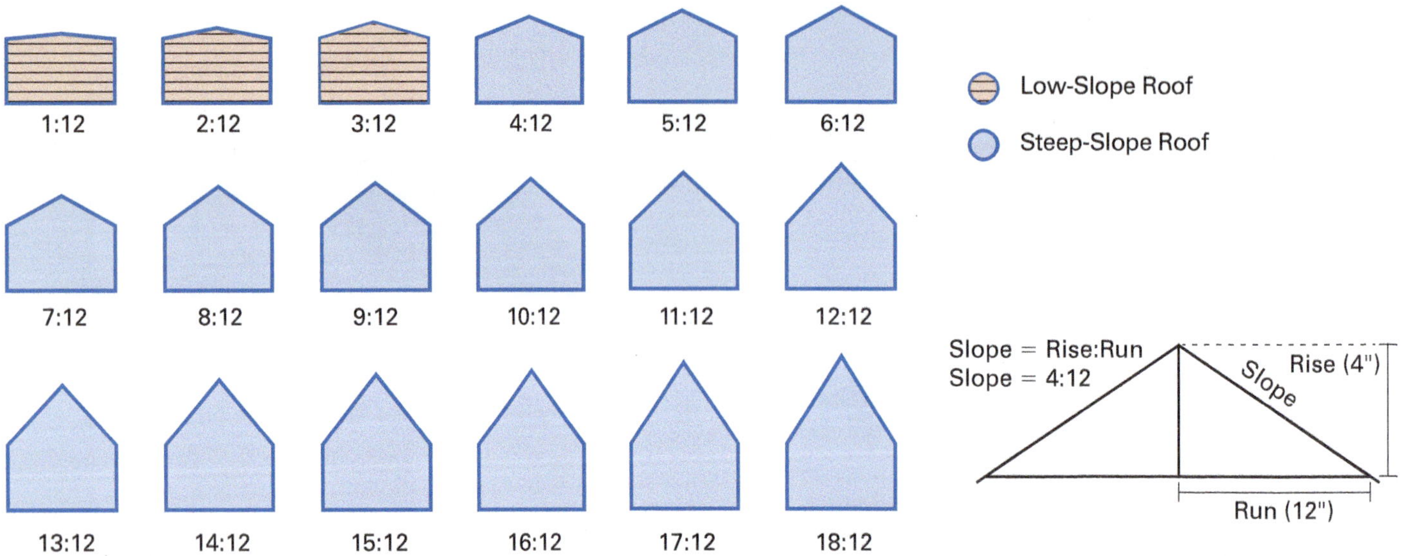

Figure 35 Calculating a roof's slope.

Figure 36 Roof-mounted options.

Source: (A) Fly View Productions/Getty Images; (B) funfunphoto/Getty Images; (C) filo/Getty Images; (D) Jason Finn/Getty Images; (E) leopictures/Shutterstock; (F) GAF; (G) Levivi/Getty Images

2.1.4 Ballasted

One type of solar PV installation option for a low-slope roof is called *ballasted*. Heavy weights, such as concrete blocks, anchor the solar PV system on the roof. This solution typically requires fewer roof penetrations, which reduces the possibility of leaks.

Note that the Department of Energy (DOE) does not recommend this type of installation in excessively high-wind regions.

2.1.5 Stand-Off Mounted (Rack Mount)

In a stand-off mounted system, the PV modules are installed three to five inches above the roof surface. Air easily flows around the PV module, which cools the PV module. As shown in *Figure 37*, there are rail and rail-less roof-mounted solutions.

(A) Rack-Mounted Solution with Rails

(B) Rail-Less, Metal, Roof-Mounted Solution

Figure 37 Rail and rail-less roof-mounted solutions.
Source: (A) krblokhin/Getty Images; (B) mikedabell/Getty Images

Solar installation materials continue to evolve, and more aesthetic integrated solutions are available, but traditionally, rack mounting has been the most common method of PV module attachment to a roof because it is robust, has versatile hardware options, and is easy to construct and install.

Roof-mounted racking systems include roof attachments, mounting rails, and module clamps. The rails may be aluminum or stainless steel. Roof attachments can be secured into concrete, metal, or wood, but each anchor point creates another roof penetration, which increases the risk of leaks. Proper weather sealing of all roof penetrations is essential. The benefit of a railed system is that the roof's rafters can be attachment points. The PV modules can be clamped anywhere along the rail to maximize roof usage.

2.1.6 Rail-Less

A rail-less system reduces the number of parts as the mounting hardware attaches the PV modules directly to the roofing surface (*Figure 36*). This system typically increases the number of roofing penetrations.

Some products have mounting hardware with pre-drilled holes for attachment to rafters and a peel-and-stick adhesive backing, so there is no chance of forgetting to seal roofing penetrations.

Not all installations can be done with a rail-less system. Some installations require anchors to be installed on the rafters to get the attachment depth required for the location's wind zone or snow load. Also, rail-less systems require specific rafter spacing and layout.

Solar PV arrays can easily be installed on standing seam metal roofs using clamps, which may allow for limited or no roof penetrations, depending upon the roof seam type.

2.1.7 Integrally Mounted

Heat fade: A condition in which a PV system operates inefficiently during periods of high heat.

Integrally mounted PV modules are built into the building or structure. They yield 10% to 15% less energy than stand-off-mounted solutions due to reduced air flow and increased heat fade. **Heat fade** is a decrease in PV module performance due to the materials overheating, and it is usually caused by poor connections or undersized wiring. Many integrally mounted solutions require a thermal underlayment.

Solar Shingles

Solar shingles are designed to be aesthetically pleasing and emulate traditional roofing shingles.

Mounting methods continue to be developed. Some shingles use common roofing nails, so the installation costs decrease. A high-temperature underlayment handles the heat produced by the solar shingles.

> **Solar Windows**
>
> Solar windows continue to advance. One product is LiquidElectricity™ liquid coating. It is applied to window surfaces and generates electricity. The material's color, tint, and amount of power are customizable.
>
> According to the Solar Window Technologies' website, skyscrapers are estimated to produce 1.3 million kilowatt-hours per year using this technology.

2.2.0 Solar Photovoltaic (PV) Cells

Solar cells are combined to make PV modules, and PV modules are grouped into solar arrays. The solar photovoltaic system comprises solar arrays, primary components, and BOS components (*Figure 38*).

Solar PV cell technology continues to develop as experimentation with different semiconductors and cell construction increases the output wattage and performance. The number of solar cells in a PV module varies by product, but historically combinations of 60, 72, 96, and 144 cells per module are common.

Doped: Having specific impurities added to a material to produce a positive or negative charge.

Most PV cells have two semiconductor layers of crystalline silicon (c-Si). The layers are chemically **doped**, which creates two distinctly charged regions

From a Solar Cell to a PV System

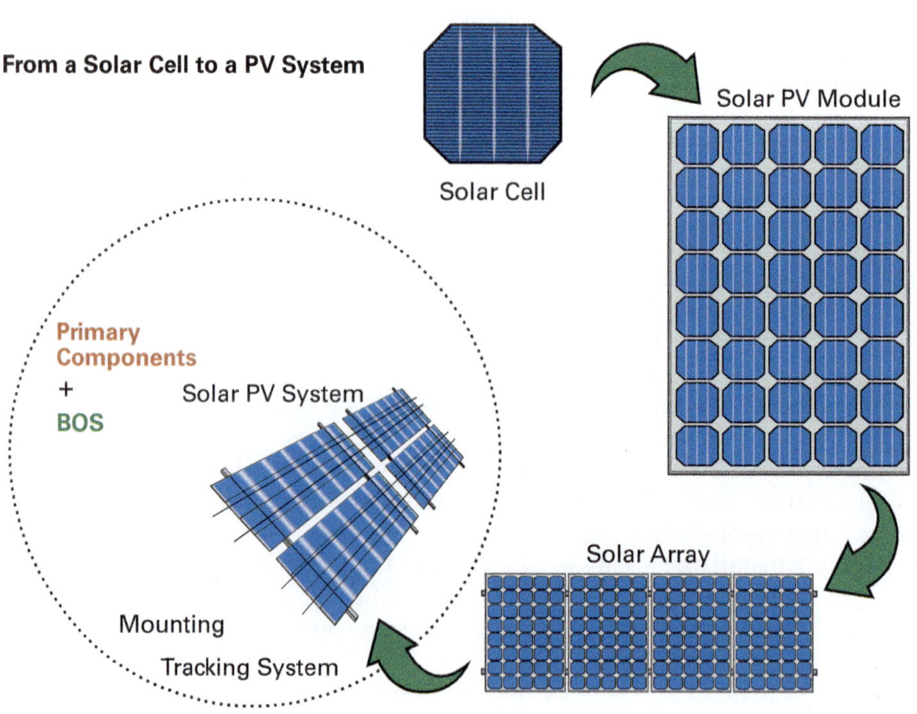

Figure 38 Solar PV system.

(positive and negative) separated by a P/N junction. These regions are sandwiched between metallic conductors.

There are three basic types of PV cells (*Figure 39*): **monocrystalline**, **polycrystalline**, and **amorphous**. **Thin film** is a flexible amorphous example. *Table 6* highlights some of the primary characteristics of each type of PV cell, such as their typical application, efficiency rating, and price. The type of silicon and manufacturing process used influences the efficiency of the solar cell.

Bifacial PV modules have solar cells on both sides of the module, which increases efficiency by absorbing reflected light. Solar cell technology continues to evolve to improve energy yield, such as absorption of reflective and ambient light.

Monocrystalline: A type of PV cell formed using thin slices of a single crystal and characterized by its high efficiency.

Polycrystalline: A type of PV cell formed by pouring liquid silicon into blocks and then slicing it into wafers. This creates non-uniform crystals with a flaked appearance that have a lower efficiency than monocrystalline cells.

Amorphous: A low-efficiency type of photovoltaic cell characterized by its ability to be used in flexible forms, like thin film.

Thin film: A low-efficiency type of photovoltaic cell characterized by its ability to be used in flexible forms. Also called *amorphous*.

NOTE

Actual PV module wattages vary widely depending on manufacturer, PV module size, and type of semiconductor. Always consult the nameplate data for the specific PV module in use.

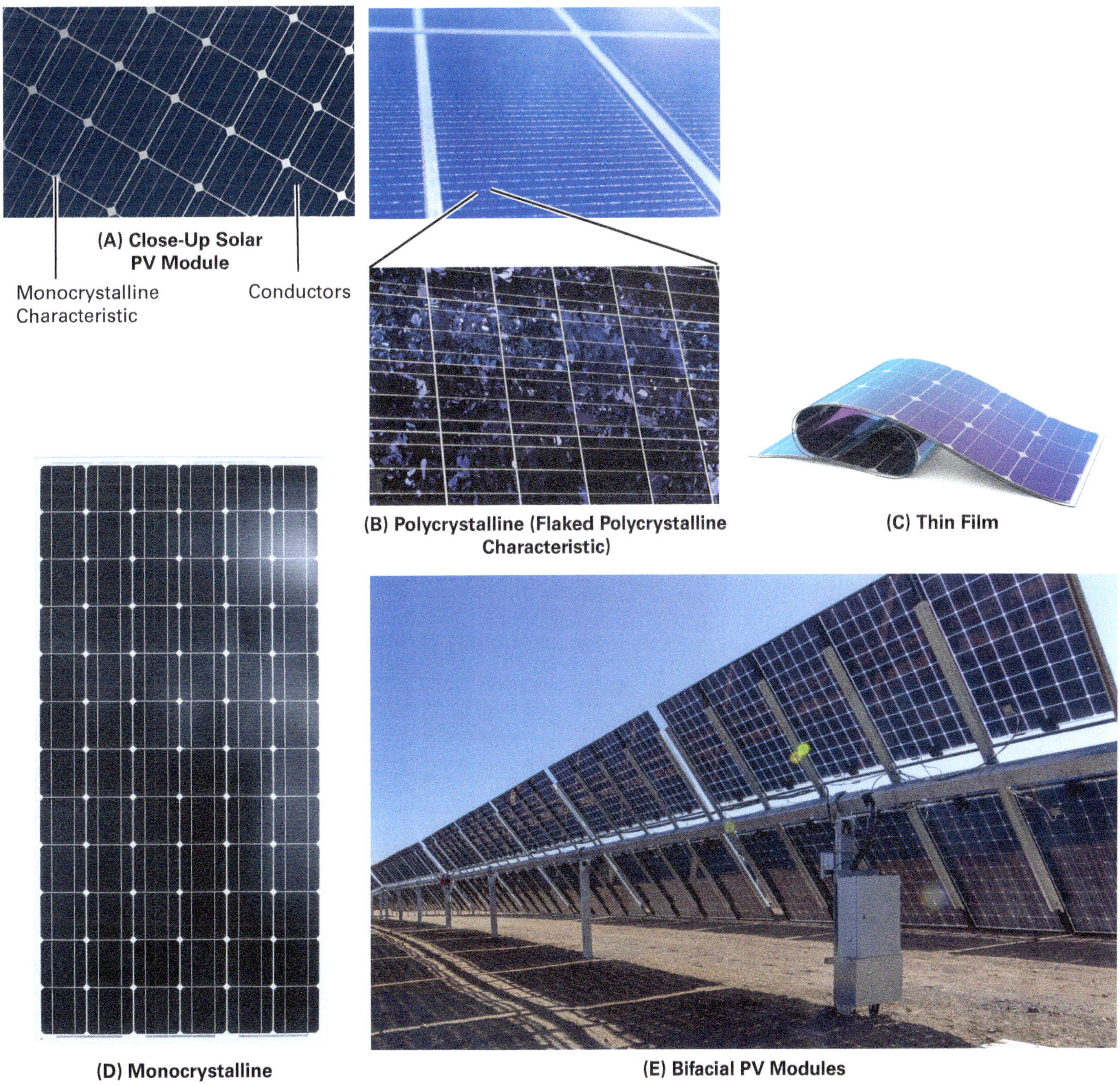

Figure 39 Types of solar PV modules.

Source: (A) Jerzy Niezabitowski/Getty Images; (B top) TomekD76/Getty Images; (B bottom) Iaremenko Sergii/Shutterstock; (C) ZU_09/Getty Images; (D) Awais khan/Shutterstock; (E) abriendomundo/Getty Images

TABLE 6 PV Cell Characteristics

	Monocrystalline Cell	Polycrystalline Cell	Thin-Film Cell
Energy	75W	65W	45W
Cost	$$$	$$	$
Application	Residential, commercial, utility scale	Residential, commercial	Residential, commercial
Key Factors	Highest efficiency rating	Poured silicon creates non-uniform crystals when it solidifies, reducing the efficiency of the PV module. Lower heat tolerance	Lowest efficiency rating Flexible material
Identifier	Older PV module: pattern of small diamonds Newer PV module: black cells	Flaked appearance	Thin-film cells are larger to produce the same amount of energy as the other types of solar cells
Manufacturing Process	Monocrystalline cells are formed using thin slices of a single silicon crystal.	Polycrystalline cells are made by pouring liquid silicon into blocks and then slicing those blocks into wafers.	Thin-film (also called amorphous) PV modules are made using ultra-thin layers of semiconductor material.

2.3.0 Solar Photovoltaic (PV) Modules

Bypass diodes: An electronic part wired in parallel with individual PV modules to direct current around a poor-performing PV module so overall array performance is less affected.

A PV module, also known as a *PV panel* (*Figure 40*), has a termination box, panel wiring in the back, and manufacturer's nameplate, or back sheet, with product specifications. Another term for termination box is *junction box*. The panel wiring has MC4 connectors that connect, or *string*, PV modules together.

PV modules can be strung in series, parallel, or series-parallel, and together, these wiring strategies help reach the maximum energy output planned for in the solar system design.

Most PV modules have **bypass diodes** in the termination boxes to prevent reverse voltage flow in the circuit. This protects cells from overheating during sustained periods of partial shading.

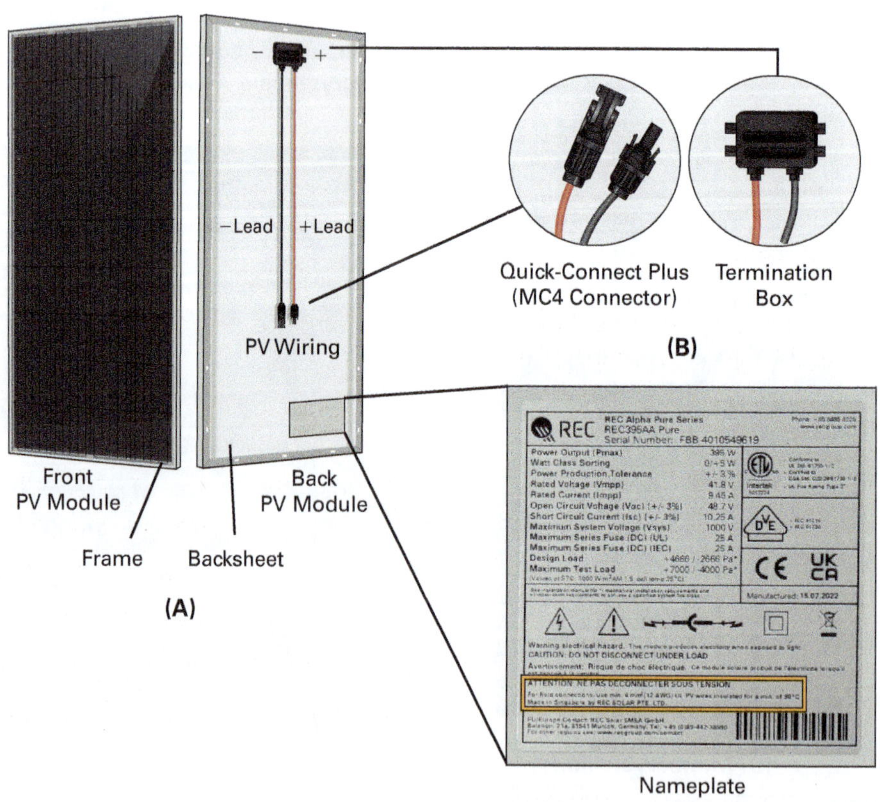

Figure 40 PV module with nameplate and panel wiring.
Source: (A) Liz Swiger, Solar Source; (B) rukawajung/Shutterstock

PV Module Size

To understand the physical space required to install a solar PV array and the amount of electricity that is produced, the following are some common 2022 PV module characteristics with variation by manufacturer:

- Produces between 150 watts and 530 watts of energy based on PV module size and efficiency
- Efficiency varies, typically between 18% and 20%
- Weighs 33 pounds to 50 pounds each
- Typical dimensions are 5.4 feet by 3.25 feet to 6.5 feet by 3.25 feet

2.3.1 Performance

PV modules are rated according to their maximum DC power output in watts under industry and internationally recognized Standard Test Conditions (STC). This information is found on the back of the PV module on its nameplate and enables the comparison of PV modules for selection for a solar PV installation.

Actual PV module performance deviates from the standard because of environmental conditions that cause system power losses or increases.

NOTE

The actual voltage output of a PV module is affected by temperature. Calculations using the STC operating temperature of 25°C are used during installation and maintenance.

Standard Test Conditions (STC)

STC testing provides a fixed set of indoor, laboratory conditions by which all PV modules can be more accurately rated and compared against each other for selection in the field.

Standard Test Conditions include the following:

- Solar cell's operating temperature of 25° Celsius (C) or 77° Fahrenheit (F)
- Incident solar irradiance level of 1,000W/m^2
- **Air mass** of less than 1.5 **spectral distribution**

Air mass: The volume of air that sunlight passes through to reach Earth.

Spectral distribution: The distortion of light through Earth's atmosphere.

Nameplate

The PV module nameplate contains system sizing and design information. It lists the model number, power output (Pmax), maximum voltage and current at Pmax, short-circuit current (I_{sc}), and open-circuit voltage (V_{oc}).

The nameplate also lists the normal operating cell temperature (NOCT) at Standard Test Conditions (STC), PV module weight, maximum system voltage, fuse rating, and recommended conductor.

As shown in *Figure 40*, this example PV module nameplate indicates 12 AWG PV wires rated for 90°C. Type USE-2 cable is commonly used.

NOTE

There are a few PV module fire rating systems which standardize the safety tests performed on PV modules. UL 790 rates the PV modules into Class A, B, and C based on the rate the flame spreads. A PV module must be installed over a roof of equivalent fire resistance.

Output Power

A PV array's actual power output depends on factors such as the PV module's efficiency, cleanliness, orientation with the Sun, amount of direct sunlight exposure, and PV module temperature. System wiring and module mismatches also contribute to system losses.

Temperature Derating

For every 1°C temperature change above and below 25°C (STC), the PV module's output voltage temporarily changes +/−0.25%. The hotter the PV module gets, the less power it generates. The colder the PV module gets, the more it generates.

Solar cell temperature variation can cause a significant change in the system output voltage. For example, looking at a PV module's nameplate, the maximum power at STC is 320W at 25°C, the open circuit voltage (V_{oc}) is 40.1V, and the temperature coefficient of maximum output power (Pmax) is −0.31%/°C. Let's say that PV module is located in Phoenix, Arizona. On a cloudless day at 120°F (48.89°C), what would the actual output be?

$$48.89°C - 25°C = 23.89°C$$

I-V Curve

The I_{SC} and V_{OC} are critical values when performing system calculations. The PV modules' current-voltage characteristics are visualized on a chart, called an I-V curve, to optimize for the maximum energy output of the PV module.

This value is multiplied by the PV module's temperature coefficient of the maximum output power (Pmax) at STC, which is found on the product's data sheet. For the example, the value is –0.31%/°C.

$$23.89°C \times -0.31\%/°C = -7.41\%$$

Because of the higher PV module temperature, this means that the PV module loses 7.41% in power output when the cells reach 48.89°C. The following equation can be used to calculate the solar module power loss.

$$-7.41\% \times 320W = -23.712W$$

$$320W - 23.712W = 296.28W$$

At 120°F (48.89°C) the maximum power this module will operate at is 296.28W.

2.4.0 Inverters

Inverters (*Figure 41*) convert the DC power produced by the PV array into AC power that can be used by various loads. There are several types: string inverter, smart inverter, microinverter, hybrid inverter, and utility-scale central inverter.

When selecting an inverter, choose a true sine wave inverter, because AC travels as a sine wave. It is more expensive than other options but has better performance.

Inverters are available in a wide variety of types and load ratings. Some inverters are designed to operate with off-grid systems, while others are designed for grid-tied and grid-interactive systems. Off-grid systems use battery-based inverter systems with an output voltage of 120/240 VAC, so they can operate larger loads, such as water heaters and stoves.

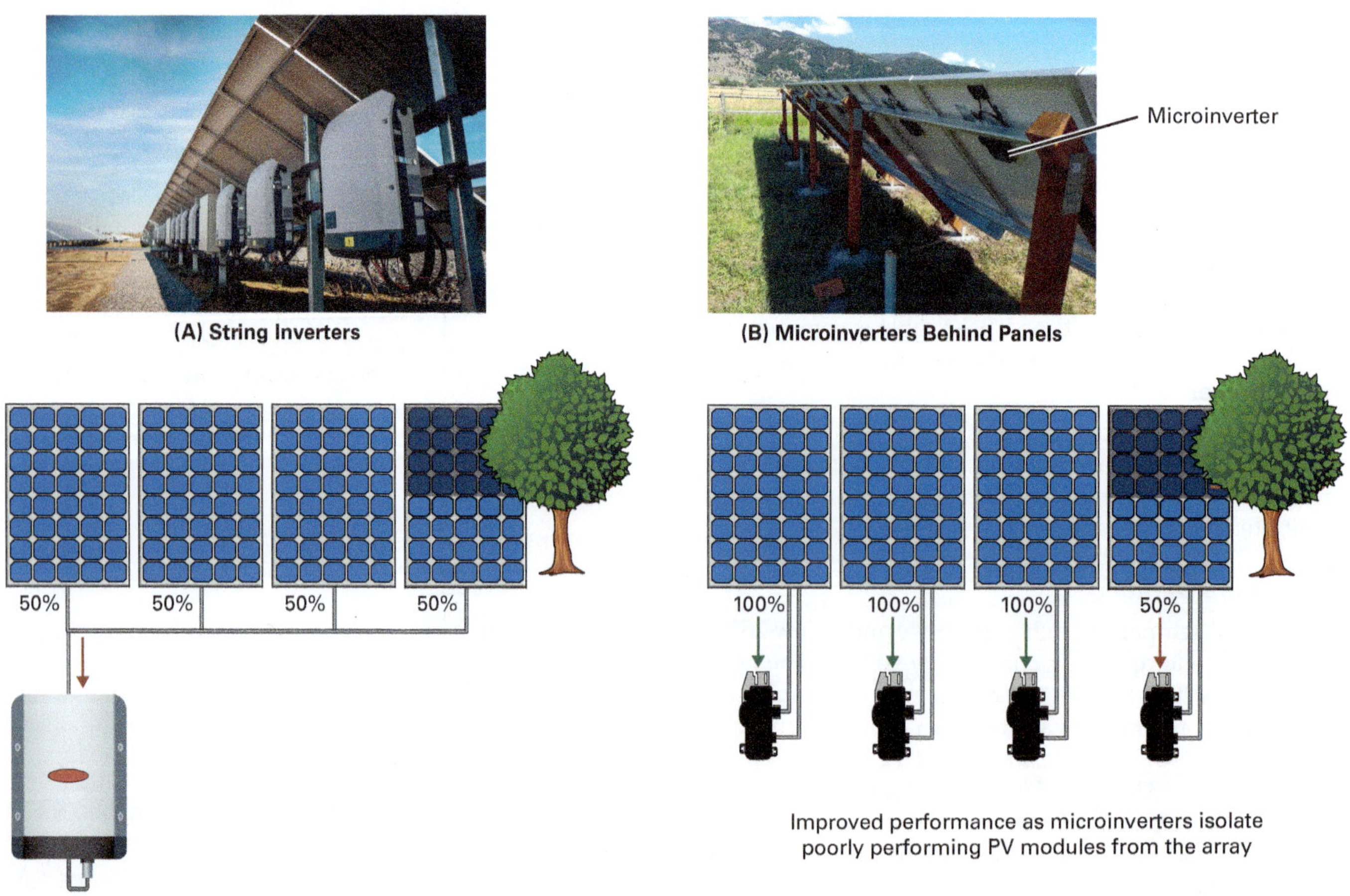

Figure 41 Types of solar inverters.
Source: (A) Roman Zaiets/Shutterstock; (B) Gary Reysa, Build It Solar

Inverters have safety ratings expressed as continuous watts and surge watts. The continuous wattage represents the total power output the inverter can support over time. The surge wattage accounts for momentary increases in wattage demand; as, for example, when an appliance's motor starts.

Some available inverter features are as follows:

- SunSpec rapid shutdown technology
- Ground-fault protection
- DC disconnect
- Arc-fault protection
- Battery charge controller
- Performance communication system with or without a digital display

2.4.1 String Inverters

String inverters tend to be centrally located and wired in series. Several strings of PV modules may feed into a single string inverter. Many inverters have a digital display to indicate **ambient temperature**, voltage output, and other system parameters. They also may have a remote monitoring system. They may need to be paired with module-level power electronics (MLPE), like an optimizer. MLPEs monitor and adjust voltage at the PV module and report PV module performance to a string inverter.

Ambient temperature: The air temperature of an environment.

Inverter Performance

Two inverter output performance indicators are compared to one another as follows:

- The estimated inverter performance, which is typically established during the design phase to understand potential system performance
- The actual inverter performance, which is measured with a category-rated voltmeter and ammeter

The estimated inverter output performance is impacted by environmental factors, including module temperature, solar irradiance, wiring losses and inverter efficiencies, and soiling. The following equation provides an estimation of the PV module's actual performance.

$$\text{Inverter output power} = \text{PV module maximum power output (STC)} \times \text{PV module temperature adjustment} \times \text{location's irradiance} \times \text{inverter efficiency} \times \text{wiring losses} \times \text{how dirty the modules are}$$

Let's work through an example (*Table 7*) for a solar array located in northern Illinois. *NEC Table 690.7(a)* contains the corrective factors to calculate the effects.

TABLE 7 Inverter Output Performance Example

Equation Variable	Where to Look Up the Value	Example
Expected PV module rating (STC)	PV module nameplate	4,800 Watts (STC)
PV module temperature adjustment	*NEC Table 690.7(a)*	A solar array with an array temperature of 30.2°F (−1°C) has a corrective factor of 1.12 $1 - [(-1°C - 25°C) \times (-0.005/°C)] = 0.87$ $1 - [(-26°C) \times (-0.005/°C)] = 0.87$ $1 - 0.13 = 0.87$
Solar irradiance	Solar irradiance map	Located in northern Illinois
Inverter efficiency	Inverter nameplate	95%
Wiring losses	Typically assumed to be 2% to 3%	3%
How dirty are the modules (*soiling*)	Typically assumed to be 2% to 5%	0% = 1.0

For example, a solar array with an array temperature of 30.2°F (–1°C) has a corrective factor of 1.12. If the array was rated to produce 4,800 watts (STC) and located in northern Illinois with a solar irradiance of 4.0, the estimated inverter output voltage can be calculated. Looking at the inverter's nameplate, its efficiency rating can be obtained.

The array temperature is 45°C and the irradiance is 840 W/m^2. The inverter efficiency is 95%; module mismatch and the DC and AC wiring losses are 2% and 3% respectively, and soiling is minimal. System adjustment factors are as follows:

Think About It

Adjust for Soiling

Recalculate the inverter output power with a soiling of 5%.

- Temperature: [1 – (45°C – 25°C) × (–0.005/°C)] = 0.90
- Irradiance: (IRR ÷ 1000W/m^2) = 840 ÷ 1000 = 0.84
- Inverter efficiency: 0.95
- Mismatch and DC and AC wire losses: 5% total = 0.95
- Soiling: 0% = 1.0

$$\text{Inverter output power} = 4800 \times (0.9) \times (0.84) \times (0.95) \times (0.95) \times (1.0) = 3275\text{W}$$

$$\text{Inverter output power} = \text{PV module maximum power output (STC)} \times \text{PV module temperature adjustment} \times \text{location's irradiance} \times \text{inverter efficiency} \times \text{wiring losses} \times \text{soiling\%}$$

2.4.2 Smart Inverters

Regulated by *Institute of Electrical and Electronics Engineers (IEEE) Standard 1547* and *Underwriters Laboratories Inc (UL) 1741*, the smart inverter, or *grid support utility interactive inverter*, has two-way communication software that shares solar performance data with the utility company's power grid. Safety features support the health of the power grid through voltage regulation to reduce exceeding grid voltage limits. The utility company has remote access to disconnect and connect the device, which categorizes the device as a point of interconnection.

2.4.3 Microinverters

Installation of microinverters is one-to-one, attaching directly to individual PV modules as they make a parallel circuit. This solution is more expensive than utilizing a string inverter. However, microinverters can ensure that the overall array performance isn't affected if, for example, one PV module in the series becomes shaded or damaged (*Figure 41*).

An *AC module* is a PV module that ships with the microinverter preinstalled as part of the product.

Microinverters have built-in rapid shutdown.

2.4.4 Hybrid Inverters

Hybrid inverters regulate charge and use of an energy storage system. They supply power from the batteries to critical loads during grid power outages. An internal switch "directs traffic" between the solar array, building, battery bank, and power grid. Depending on the product features, this device may not replace the main solar inverter but may be an additional component in the system.

2.4.5 Utility-Scale Inverters

Utility-scale projects use central inverters with larger capacities from 500kW to 800kW. Decentralized systems (e.g., string inverters) are becoming more popular where larger-rated three-phase commercial inverters rated from 50kW to 250kW per inverter are used. These inverters are combined in large combiner boxes or at a centralized switchgear system.

For grid connection, some inverters' low-voltage output (typically 300V to 450V) is stepped up with two sets of transformers (typically 11kV and above) before the interconnection point with the power grid. Voltages for large commercial and utility-scale systems now range from 1,000V to 1,500V.

2.5.0 Energy Storage Solutions

Solar energy can be stored in battery banks with charge controllers or smart inverters to regulate the DC electricity flow in and out of the batteries. A solar sensor opens and closes the circuit, drawing power from the battery at night, for example.

Residential, commercial, and utility-scale energy storage solutions can be used for the following purposes:

- Supply startup currents to electrical loads, such as motors, when the solar array isn't generating electricity
- Operate the PV array near its maximum power point, to stabilize output voltages during periods of low production
- Charge other batteries, such as an electric vehicle's battery

2.5.1 Batteries

Batteries store and discharge DC power. In a process called cycling, they discharge, or release energy, and then can be recharged. Distinguishing battery characteristics include voltage (V), amp-hours (Ah), deep cycling, depth of discharge, sealed, and maintenance free. Batteries are typically 12V, 24V, or 48V with safety recommendations to limit the number of parallel strings to no more than four. Batteries have their voltage and amp-hours printed on the case.

Lithium-ion batteries lead the market because they offer the following advantages:

- Do not vent hydrogen gas
- High density, which means lots of energy storage in a small space
- Deep cycling with high depth of discharge
- Maintenance free
- Common energy storage solution for electric vehicles (EV) and electronics

Maintenance free means that the battery does not require electrolyte inspection or periodic refilling with distilled water.

Table 8 lists common types of batteries.

Bluetooth features are included in some batteries to report battery status to smartphone applications.

Lead-Acid Batteries

Historically, lead-acid batteries were used in cars and lent themselves to early solar PV system designs. Examples of lead-acid batteries include conventional and sealed, maintenance free.

A lead-acid battery (*Figure 42*) has cells that contain a sulfuric acid electrolyte (liquid), which creates a chemical reaction to store energy. Each cell produces 2V, so a 6V battery contains three cells and a 12V contains six cells.

Conventional batteries, like flooded lead-acid (FLA) batteries, require regular maintenance as the electrolyte boils off. If the liquid level is low, fill the cells with distilled water through the filler plug. Dry cells lead to sulfation, which damages the battery. The chemical reaction within the cells outgasses dangerous hydrogen gas. For safety, store FLA batteries in vented, high-density polyethylene (HDPE) containers. Maintenance of the terminals is also important to prevent corrosion.

WARNING!

Overcharging and improper venting of FLA batteries can cause an explosive gas buildup within a battery. When this occurs, the battery may leak, feel warm, or exhibit signs of swelling. The pressure rises within the cells until a nearby spark or short circuit ignites the gas. Battery explosions release violent sprays of caustic acid and shrapnel that can cause severe injury or even death.

TABLE 8 Battery Types

Battery Type	Description	Sector	Application Notes
Lithium-ion	Deep cycle, high-density battery with 95% depth of charge	Residential, large scale	Great for off-grid and grid-tied systems; lifespan twice as long as sealed lead-acid (SLA) batteries Maintenance free Sold as electric vehicle battery solution
Gel, SLA	Deep cycle, hybrid gel	Residential	Maintenance free
SLA	Deep cycle, absorbent glass mat (AGM) battery with 50% depth of charge	Residential	Great for off-grid systems and during emergencies Maintenance free
Lead-acid	Flooded lead-acid (FLA) battery with 50% depth of charge	Residential	Great for off-grid and during emergencies Requires refills, so needs maintenance access, and space venting for hydrogen outgassing
Nickel-cadmium	Varies by battery size; large capacities and high discharge rates	Residential, large scale	Withstands extreme temperatures Durable, maintenance free Difficult to dispose of; banned in some countries
Zinc-bromine flow battery	Deep cycle (5,000+), potential replacement for lithium-ion batteries with lower risk of fire due to nonflammable electrolyte solution	Large scale	Capable of 100% depth of discharge Does not require a cooling system

Batteries evolved with sealed, maintenance free, and gel options to minimize safety hazards and remove maintenance requirements. For example, absorbed glass mat (AGM) batteries generally do not leak or produce corrosive gases.

Industrial-scale lead-acid batteries can last as long as 20 years with little maintenance.

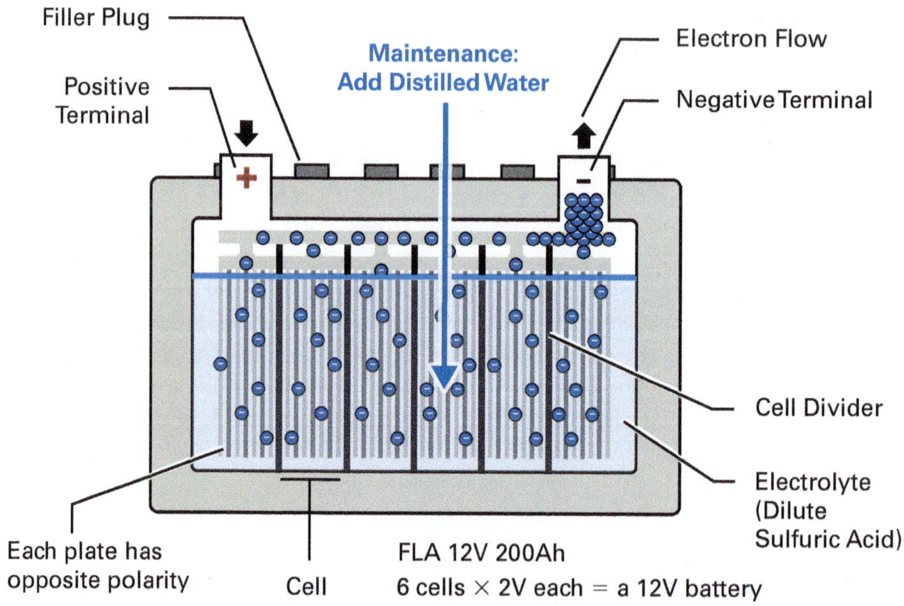

Figure 42 Lead-acid battery.

Battery Life

Batteries are rated by the expected number of charge cycles, as well as the maximum discharge. Battery life depends on the number of cycles and depth of discharge per cycle. Some PV systems include a low battery warning light or cutoff switch to prevent the damage caused by repeated deep cycling.

Batteries store energy, but slowly discharge 100% of that energy over time if they aren't recharged with more energy. The amount of charge drained is known as the **depth of discharge (DOD)**. Battery capacity is measured in amp-hours (Ah). For example, a 12V 200Ah battery from which 160Ah of energy has been used has undergone an 80% DOD.

Depth of discharge (DOD): The percentage of a battery's capacity that may be safely used to power a load without damaging the battery.

A deep-cycle battery is designed to regularly discharge most, but not all, of its capacity. For example, lithium-ion batteries may be discharged to 95% capacity (*Figure 43*).

While sealed lead-acid, deep-cycle batteries can be discharged by 80%, it is best not to go that low as repeated deep cycling shortens battery life. When sizing the battery bank, a conservative approach is taken by padding the manufacturer ratings. The solar design typically allows for 80% discharge for lithium-ion batteries and 50% discharge for sealed lead-acid.

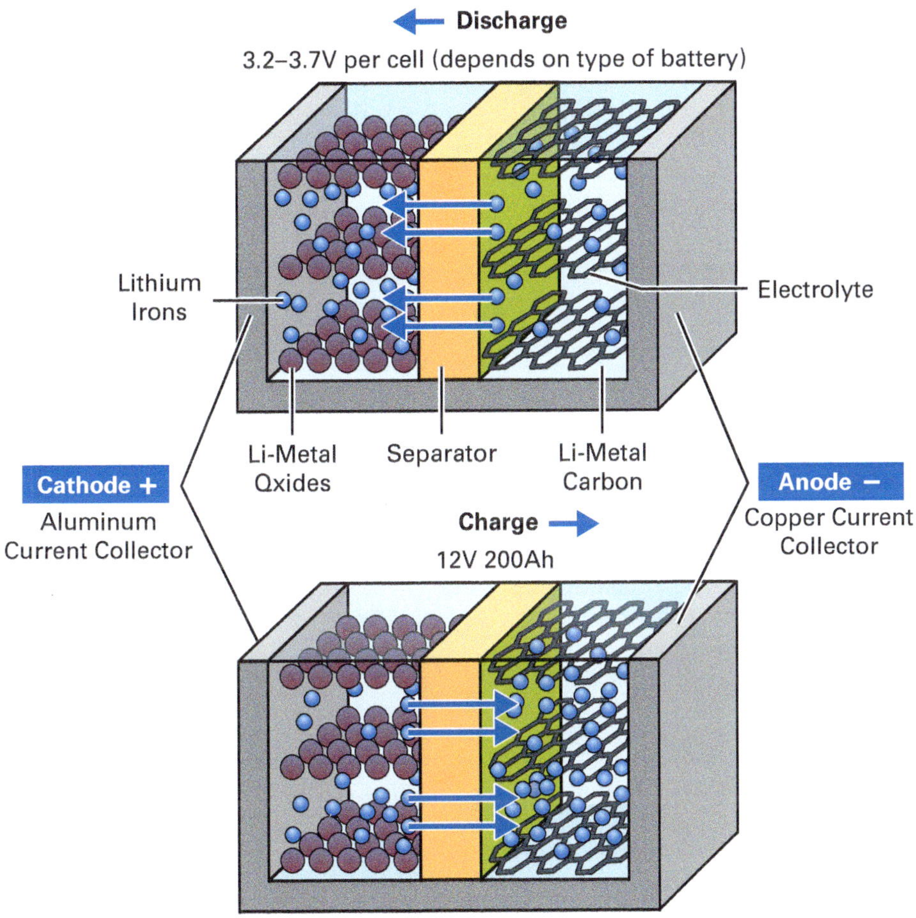

Figure 43 Lithium-ion battery.
Source: https://www.shutterstock.com/image-illustration/how-lithium-ion-battery-works-3d-1527294641

Maintenance

NEC Articles 480 and *706.20* provide crucial details on installation and maintenance information such as support, location, ventilation, and labeling requirements for stationary storage battery systems and energy storage systems (ESS).

Protect batteries from extreme temperatures. Cold temperatures reduce battery output while hot temperatures increase deterioration. With proper care and appropriate cycling, batteries can last 5 to 15 years.

Electrical Vehicle (EV) Charging

It is projected that by 2030, electric vehicle sales will dominate the market as it responds to US policies to tackle climate change.

Electric vehicles require DC power to charge their lithium-ion batteries. There are three common power sources (*Table 9*) for EV charging that use the car's charging port and onboard charger. Connections vary by manufacturer, but any AC power is converted to DC prior to charging the battery.

TABLE 9 Charging an Electric Vehicle

Power Source	Requirements	Charge Rate
Standard 120V outlet	Requires a portable charger	Overnight for 40 miles
240V outlet	Requires charging station installation	Less than 2 hours for 40 miles
Public, DC fast charging station	Requires specific charging port	1 minute for 10–20 miles

An electric vehicle (EV) battery may be included in a solar PV design as an energy storage solution. Vehicle-to-grid (V2G) or vehicle-to-home (V2H) is a technology that enables energy to be transferred from the battery of an EV to the home loads during an off-grid scenario.

Alternatively, through agreements with local electrical utilities, it may be possible for power companies to access EV batteries and use them as a grid asset, to offset loads during peak energy demand hours. The exact framework for V2G interconnection(s) is still in development.

Vehicle-to-everything (V2X) enables an EV battery to transfer energy to a home or industrial building. *Figure 44* is an example of a grid-interactive configuration with V2X. In most cases, however, this type of configuration uses a lithium-ion battery bank solution.

Figure 44 PV components with EV charging station.
Source: Petmal/Getty Images

Utility-Scale Applications

Vehicle charging and the general increase in electricity demands on the power grid will require new battery technologies to increase battery capacities at utility scale.

Nickel-cadmium and zinc-bromine flow batteries are found at the utility scale. They are durable, capable of operating at extreme temperatures, and made of extremely toxic materials. Nickel-cadmium batteries require very little maintenance.

2.5.2 Charge Controllers

Charge controllers regulate the charge and discharge of the system batteries. They are rated by their maximum AC current output, so it's essential that they be sized to match the application. Sometimes they are integrated into the inverter.

Two main types of charge controllers are **pulse-width modulation (PWM)** and **maximum power point tracking (MPPT)**. PWM charge controllers are ideal for use with sealed batteries but can only be used within a limited range of PV

Pulse-width modulation (PWM): A method of turning off and on the power in a circuit to decrease current.

Maximum power point tracking (MPPT): A battery charge controller that provides precise charge/discharge control over a wide range of temperatures.

module configurations and voltages. The PWM charge control cycles the power off and on, decreasing the current to the batteries to slow the charge rate for safety reasons and to maintain the life of the battery.

WARNING!

Older charge controllers use shunts or relay transistors and do not work well with sealed batteries.

The MPPT charge controller (*Figure 45*) is the most versatile, efficient, and costly option. A digital display shows power use, battery charge, and other system performance data. MPPT charge controllers are preferred for use in cold climates because they prevent the overvoltage that can occur at lower temperatures and provide more precise control over the battery's depth of discharge (DOD).

Figure 45 MPPT charge controller.
Source: Anthony Johnson

2.0.0 Section Review Questions

1. Which type of installation requires soil studies to guide foundational design decisions?
   a. Building-integrated system
   b. Ground-mounted system
   c. Roof-mounted system
   d. Stand-off mounted system

2. The NCCER Roofing Program defines a steep-slope roof as having a minimum rise-to-run ratio of _____.
   a. 3:12
   b. 4:12
   c. 8:12
   d. 16:12

3. A type of PV module having a characteristic flaked appearance is probably made up of _____.
   a. monocrystalline cells
   b. megacrystalline cells
   c. polycrystalline cells
   d. amorphous cells

4. Industry and internationally recognized Standard Test Conditions (STC) rate PV modules according to what characteristic?
   a. Maximum AC power output in watts
   b. Maximum DC power output in watts
   c. Maximum system voltage in volts
   d. Short-circuit current (I_{SC})

5. What is the main benefit of installing a microinverter instead of a string inverter?
   a. AC power is inverted to DC power.
   b. DC power is inverted to AC power.
   c. Poor performing PV modules are isolated from the PV array.
   d. Installing a microinverter is less expensive.

6. Which of the following batteries must be filled periodically with distilled water?
   a. Flooded lead-acid (FLA) battery
   b. Lithium-ion battery
   c. Sealed lead-acid (SLA) battery
   d. Zinc-bromine flow battery

3.0.0 Balance of System (BOS) Components

Performance Tasks

This is a knowledge-based module. There are no Performance Tasks.

Objective

Identify a PV system's balance of system (BOS) components.

a. Identify mechanical BOS components within the system.

b. Identify electrical BOS components within the system.

Solar system designers plan for the primary components, which include PV modules, inverters, controllers, and interconnection system equipment. To execute that plan, balance of system (BOS) components, such as wire and fasteners, are required. The term *BOS components* covers all the additional mechanical and electrical parts required to connect those primary components. Typical BOS components are as follows:

- Safety components: grounding, overprotection devices, rapid shutdown devices, and disconnect switches
- Raceways and conduits
- Junction and combiner boxes
- Disconnect switches
- Fuses and circuit breakers
- Terminals and connectors
- Array mounting and other structural hardware (i.e., fasteners, clamps)

3.1.0 Mechanical BOS Components

NOTE

To learn more about the assembly of systems, refer to NCCER Module 57104, *System Installation and Inspection*.

The mechanical BOS components support the primary components based on the engineer's specifications outlined in the solar design.

Clamps, screws, concrete footings versus pier footings, racking hardware, cable trays versus conduit, and all the other parts that complete the installation of the solar PV primary components are detailed by the engineer in the solar design. The engineer makes decisions regarding which parts to include, and performs a cost analysis of parts. For example, when compared to conduit, it may be cheaper to use cable trays to run a high volume of wiring on a commercial rooftop. Choosing a racking system is another example. The assembly times and associated labor costs of various systems are compared and factored into the decision-making process.

3.2.0 Electrical System Components

Electrical system BOS components include the PV module wiring, combiner boxes, disconnect switches, grounding, and termination boxes. All wiring must be sized for the load and the distance between components. Unless the cable is classified as low voltage, it must be routed in conduit.

WARNING!

Wiring at the termination box must be completed by a certified electrician.

3.2.1 Wire

Wires include a conductor, which carries the current, expressed in *amperes*. They are wrapped in a color-coded insulation that designates their function and contains the electrons along the path of the circuit.

Copper's chemical properties (e.g., high heat resistance, electrical conductivity, and malleability) make it a great conductor, so it is commonly used in wiring. The ampacity and resistance ratings shown in *Table 10* are for copper wire.

TABLE 10 Wiring

Represention of Wire Size	AWG	Ampacity			Resistance	Use
		60°C	75°C	90°C	Ω/ft	
						Power over ethernet (PoE), LED lighting
18 AWG 	18	10	14	16	0.00795	Low-voltage systems
16 AWG 	16			18	0.00499	
14 AWG 	14	15	20	25	0.00314	Small appliances and lighting
12 AWG 	12	20	25	30	0.00198	Small appliances and lighting
10 AWG 	10	30	35	40	0.00124	Appliances (120V)
8 AWG 	8	40	50	55	0.000778	Appliances (120V)
6 AWG 	6	55	65	75	0.000491	Appliances (240V), grounding
4 AWG 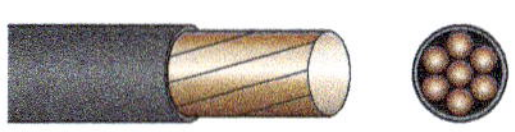	4	70	85	90	0.000308	Appliances (240V), grounding
3 AWG 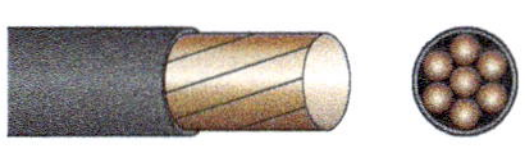	3	85	100	115	0.000245	
2 AWG 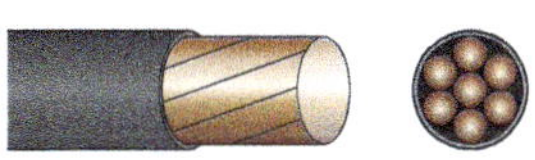	2	95	115	130	0.000194	Service entrance use, grounding
1 AWG 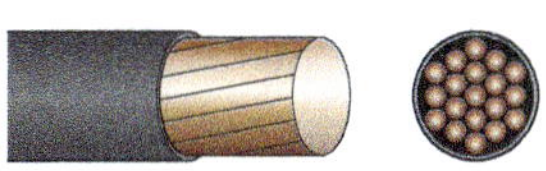	1	110	130	145	0.000154	Service entrance use, grounding
0 AWG 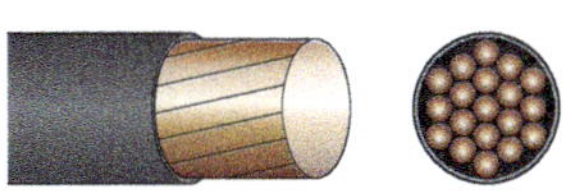	1/0	125	150	170		Service entrance use
2/0 AWG 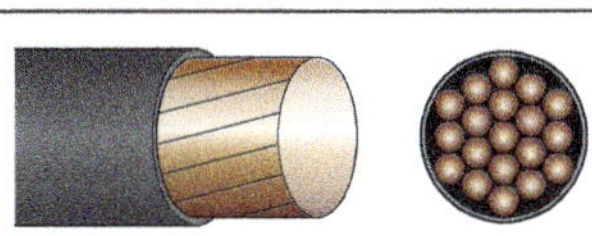	2/0	145	175	195		Service entrance use
3/0 AWG 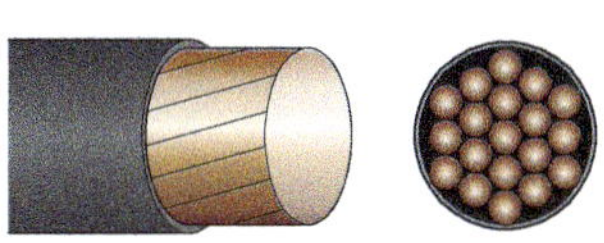	3/0	165	200	225		Service entrance use
4/0 AWG 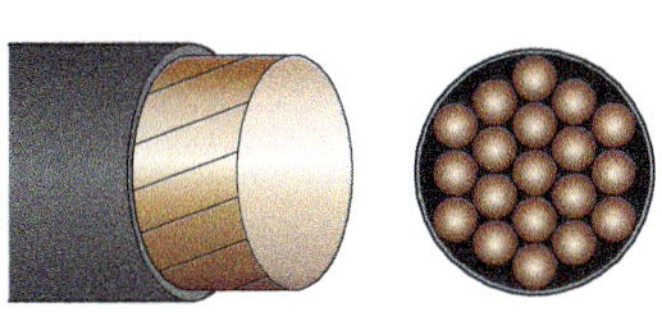	4/0	195	230	265		Service entrance use

Did You Know?

Aluminum Wire

Aluminum wire is used in overhead transmission and distribution lines by utility companies. It is less expensive and has less conductivity than copper. It has a lower melting point. In situations where aluminum has been used in place of copper, the AWG is increased by two wire sizes.

NOTE

The standard system of wire sizes in the United States is the American Wire Gauge (AWG) system. *NEC Chapter 9, Table 8* has information on wire sizes. Use *NEC Table 310.104(A)* to identify types of insulation that are suitable for use in wet locations.

Wire is sized by the American Wire Gauge (AWG) system. Wire gauge refers to the physical size of the wire, rated with a number designation that runs opposite to the diameter of the conductors. Common sizes include 16-, 14-, 12-, 10-, 8-, 6-, and 2-gauge wire. The size of the wire dictates how much current can safely pass through the wire. The smaller the wire gauge number, the larger the wire diameter. So, 10 AWG wire has a larger diameter than 12 AWG wire.

The more current the wire needs to carry, the larger the wire needs to be, but it must remain flexible. For this reason, wire sizes of 6 AWG or larger are stranded. Stranded wires use several wires, bundled together, to carry the current.

The wire size, purpose, and voltage rating are marked on the insulation (*Figure 46*).

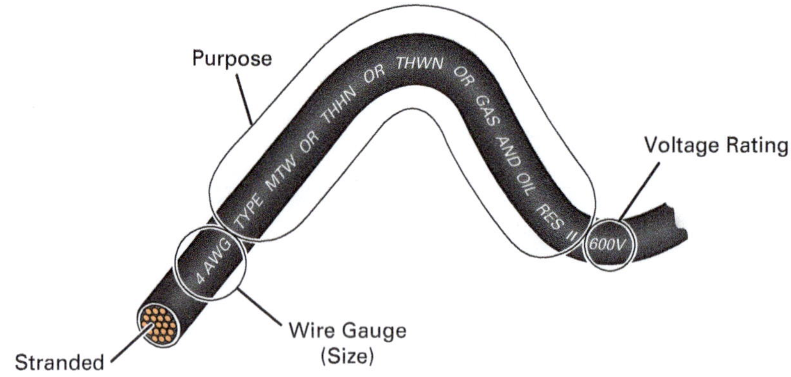

Figure 46 Wire insulation markings.

Color Coding

A color code is used to help identify wires by the color of the insulation. This makes it easier to select and properly connect the wires. The common standards for wire colors are listed in *Table 11*.

The insulation is stamped with a code that indicates the purpose of the wire. *Table 12* lists common insulation codes used in solar PV installations.

TABLE 11 US AC Power Circuit Wire Insulation Color Codes

Function	Label	Color, Common	Color, Alternative
Protective ground	PG	bare, green, or green-yellow	green
Neutral	N	white	grey
Line, single phase	L	black or red (2nd hot)	
Line, three phase	L1	black	brown
Line, three phase	L2	red	orange
Line, three phase	L3	blue	yellow

TABLE 12 Insulation Coding

Letter	Description
THHN	Heat-resistant thermoplastic
THHW	Moisture- and heat-resistant thermoplastic
THW	Moisture- and heat-resistant thermoplastic
THWN	Weather-rated, heat-rated thermoplastic with nylon cover
UF	Underground feeder
USE	Underground service entrance
XHH	Thermoset
XHHW	Heat-rated, flame-retardant, weather-rated thermoset

Voltage Drop

Minimizing the length of wire between PV components reduces voltage drop. Voltage drop is the voltage loss in the system due to conductor resistance. Any loss of power in the system reduces performance. The acceptable, or maximum, voltage drop in a solar system design is 3% (typically, 2% on the DC side and 1% on the AC side).

For each conductor run, for example, the DC side may be from the array to the combiner box, from the combiner box to the DC disconnect, and from the DC disconnect to the DC input on the inverter, which requires the summation of the three voltage drop calculations.

The basic formula for calculating voltage drop (Vdrop) is as follows:

$$\text{Voltage drop \%} = (I \times R \times N \times L)/\text{Vmpv}$$

where:

I = the operating current
R = the wire's resistance or reactance (Ω per 1000 ft)/1000 ft
N = phase coefficient (2 for single phase, 1.73 for three phase)
L = conductor length from point to point (i.e., the array to the main electrical service panel)
Vmpv = maximum power voltage

Using the PV module's nameplate values and *NEC®* lookup tables for conductor resistance, a solar system designer can estimate voltage drop within the system and keep within acceptable limits through selection of wire or placing BOS components close to each other as shown in the following example.

Example: Can you use 4 AWG wire with a 38.9A system that is 300 ft away from the point of interconnection on a three-phase AC, 240V circuit?

I = 38.9A

R = 0.314 ohm/1000 ft, based on *NEC Chapter 9, Table 9* for AC and *NEC Chapter 9, Table 8* for DC

N = 1.73 for three phase

L = 300 ft

Vmpv = 240V

$$\text{Vdrop} = (1.73 \times 0.314\ \text{ohm} \times 38.9\text{A} \times 300\ \text{ft})/1000\ \text{ft}$$

$$\text{Vdrop} = 6.34\text{V}$$

$$\text{\%Vdrop} = (6.34\text{V}/240\text{V}) \times 100 = 2.64\%$$

Because 2.64% is less than 3%, 4 AWG wire is acceptable.

Selecting Wire

Five factors are considered when selecting wire: capacity of amperage, insulation type, how flexible the wire needs to be, insulation color, and conductor material (copper versus aluminum).

A critical factor in selecting conductors is the conductor's maximum operating temperature. Consider how and where a conductor will be used so that the conductor's limiting (maximum) temperature rating will not be exceeded.

Wiring decisions are based on details about the installation, such as the temperature rating of the termination, the ambient (surrounding) temperature, the number of conductors in a bundle, and where the conductors are installed (raceways, conduit, ducts, underground, etc.).

Maximum safe carrying capacity is determined by the size of the wire. As *Table 10* indicates, the smaller the wire gauge, the higher the resistance.

Conductor selection is based largely on the temperature rating of the wire. This requirement is extremely important and is the basis of safe operation for insulated conductors. As shown in *NEC Table 310.104(A)*, conductors have various temperature ratings. Because *NEC Tables 310.15(B)(16)* and *310.15(B)(17)* are based on an assumed ambient temperature of 30°C (86°F), conductor ampacities are based on the ambient temperature plus the heat (I^2R) produced by the conductor while carrying current. Therefore, the type of insulation used on the conductor is the first consideration in determining the maximum permitted conductor ampacity.

WARNING!

Color designations are good indicators, but never trust your life to them. Always protect yourself by testing circuits with a voltmeter or multimeter rated for high-voltage systems.

Choosing the type of wire insulation is based on the following factors:

- *Voltage* — the higher the voltage level, the thicker the insulation needs to be to reduce the risk of shock
- *Environmental considerations* — exposure to the Sun and UV light, extreme temperatures, presence of chemicals or oils
- *Abrasion resistance* — higher vibration environments require thicker insulation
- *Coating* — some coatings can make it easier to push/pull conductors through conduit

NOTE

When more than three conductors are used in a conduit or raceway, use a 90-degree table.

For example, a 3/0 THW copper conductor for use in a raceway has an ampacity of 200A according to *NEC Table 310.15(B)(16)*. In a 30°C (86°F) ambient temperature, the conductor is subjected to this temperature when it carries no current. A THW-insulated conductor is rated at 75°C (167°F), which leaves a wide margin for increased temperature due to current flow. If the ambient temperature exceeds 30°C (86°F), the conductor maximum load-current rating must be reduced proportionally per *NEC Table 310.15(B)(2)(a)* so that the total temperature (ambient plus conductor temperature rise due to current flow) will not exceed the temperature rating of the conductor insulation. For the same reason, the allowable ampacity must be reduced when more than three conductors are contained in a raceway or cable. See *NEC Section 310.15(B)(3)(a)*.

Using the ampacity tables: An important step in circuit design is the selection of the type of conductor to be used (TW, THW, THWN, RHH, THHN, XHHW, etc.). The various types of conductors are covered in *NEC Article 310*, including their ampacities for the varying conditions of use (e.g., in a raceway, in open air, at normal or higher-than-normal ambient temperatures). Conductors must be used in accordance with the data in these tables and notes.

3.2.2 Grounding and Bonding

PV components are grounded. To ground the system, grounding wires connect to grounding rods to discharge energy to the earth, literally to ground. Grounding occurs at two places within the PV system. One is located on the DC side near the PV modules and the other is on the AC side near the main electrical service panel. For a roof installation, 6 AWG copper wire is the minimum size wire gauge that may be used for the roofing grounding wire.

The PV components are bonded. This means all exposed noncurrent-carrying metal parts of the equipment are electrically connected to one another, which connects into the grounding system. The ground connection must be continuous.

CAUTION

PV modules have a grounding wire that runs along the frame, which should be installed before connecting the PV modules.

In the event of a short circuit, the ground takes excess electricity and "runs it to ground," which has zero voltage. This safety measure reduces the risk of electrical shock.

PV modules are no exception. Manufacturers may provide hardware for grounding and bonding the PV modules and rails to one another. This hardware may include lugs, grounding clips, self-grounding clamps, bonding splice bars, and bonding jumpers. If the manufacturer hasn't provided the hardware, you need an electrically sound plan with noncorrosive materials. Long installations require joining multiple rails together. The rails are spliced together, and lugs are attached from rail to splice to next rail to maintain the bonding.

The system size, soil composition, water levels, weather patterns, depth of ground freezing, and type of racking will all factor into how the system is grounded.

Utility Scale

Each site is unique and will have different requirements. In general, a grounding grid is a network of conductors placed in the soil beneath the substation. It is bonded to all metal noncurrent-carrying equipment and structures, such as equipment housings and fences. The purpose of the grounding grid is to safely route any high voltages that result from equipment switching, malfunctions, or lightning strikes to earth, literally ground.

Additional Safety Measures

Arc-fault protection is required per *NEC Section 690.11*. Ground-fault protection is required per *NEC Section 690.9*. Some inverters include integral ground-fault protection. If it is not supplied in the inverter, it must be located elsewhere in the PV system.

Lightning Arrestors

Lightning arrestors are installed in addition to grounding rods and may include surge protectors, which redirect excess current to ground as needed.

3.2.3 Overcurrent Protection

Overcurrent protection includes circuit safety devices, such as fuses and circuit breakers.

PV components, including inverters and charge controllers, have limits to the amount of current they can safely handle. Codes and standards can be found in the *NEC*® guidelines. Overcurrent protection opens, or *breaks*, the circuit to stop the unsafe flow of electricity, which could potentially lead to a fire.

Fuses are intended for one-time use. They are commonly referred to as *blown* if the current exceeds the fuse's capacity, which safely melts the fuse's internal wiring at high temperature.

Circuit breakers are thrown, or *opened*, breaking the circuit, using an internal electromagnetic switch that flips the breaker to the Off position. An electrician fixes any wiring problem and then flips the circuit breaker switch to the On position to reconnect, or *close*, the circuit.

NOTE

To learn more about overcurrent protection, including how to size the fuse for the application, refer to NCCER Module 57104, *System Installation and Inspection*.

3.2.4 Combiner Box

A combiner box is an electrical distribution box with DC safety overprotection used to combine multiple sets of PV module strings into one set of conductors via a fused terminal block. This reduces the number of conductors for a long run and upsizes the wire to manage voltage drop. Combiner boxes may include a fused DC disconnect.

A **home run** is the main collection of wires that connect a fixture to a centralized location for power distribution. In the case of the combiner box, multiple home runs from the array are reduced to one wire that leads to the inverter.

Home run: Main collection of wires that connect a fixture to a centralized location for power distribution.

3.2.5 Rapid Shutdown and Disconnect Switches

A PV Hazard Control System (PVHCS), commonly referred to as a *rapid shutdown device* or system, is an electronic safety device that reduces voltage output levels to less than or equal to 80V within 30 seconds of initiating a rapid shutdown. This safety feature reduces the electrical hazard risk for firefighters during an emergency and is covered by the safety standard *UL 3741*. Rapid shutdown guidelines are detailed in *NEC Section 690.12*.

Disconnect switches must be labeled and installed at ground level for access during an emergency. Exact locations vary by municipality, so always follow engineering and Authority Having Jurisdiction (AHJ) guidelines.

The DC and AC sides of the PV system both require a disconnect switch, also known as a *breaker*, to disconnect the power from that part of the PV system. Disconnecting means are detailed in *NEC Section 690.13* and *NEC Section 690.15*.

Example AC and DC disconnects are shown in *Figure 47*. The DC disconnect is located on the PV module side of the inverter. Some inverters and combiner boxes include a DC disconnect.

The AC disconnect, or rapid shutdown, is located near the service panel. PV disconnects may be located indoors if the panel is in the building.

NOTE

A combiner box is a type of junction box, but not all junction boxes are combiner boxes. Also, not all combiner boxes are weatherproof.

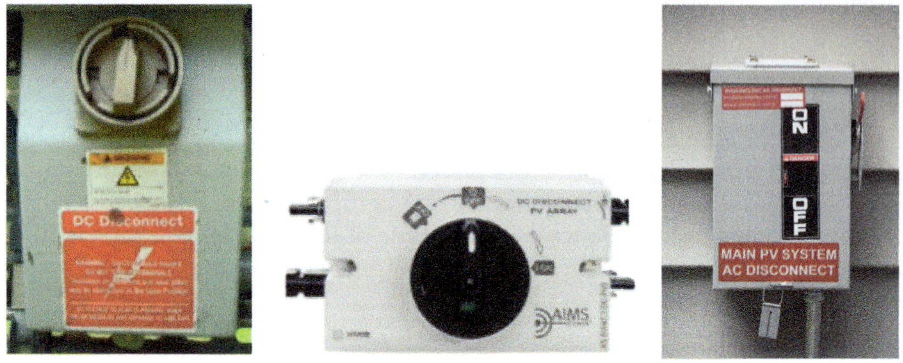

(A) Combiner Box with DC Disconnect **(B) DC Disconnects** **(C) AC Disconnect**

Figure 47 Types of DC and AC disconnects.
Source: (C) filo/Getty Images

A wire exits the combiner box connecting to a DC disconnect. This disconnect is a switch that opens the circuit when flipped to the Off position, isolating the PV modules' DC power from the rest of the circuit and the building.

An AC disconnect is a switch that opens the circuit when flipped to the Off position, isolating the building from the solar PV system.

WARNING!

Electrical shock hazard to firefighters is high as high voltage remains in the line for at least 5 minutes after the rapid shutdown switch has been thrown. The DC disconnect switch does not stop the PV modules from producing electricity; instead, it stops the flow of that electricity within the system. Apply warning labels as required.

3.2.6 Switchgear

Switchgear

Switchgear often utilizes draw-out breakers, which can be easily removed for maintenance or replacement without shutting down the whole system.

Switchgear provides overprotection and safety disconnects. It looks like a large-scale electrical panel but is rated for up to 38kV with a current rating of up to 6,000A. The *NEC®* identifies switchgear as "an assembly completely enclosed on all sides and top with sheet metal (except for ventilating openings and inspection windows) and containing primary power circuit switching, interrupting devices, or both, with buses and connections. The assembly may include control and auxiliary devices. Access to the interior of the enclosure is provided by doors, removable covers, or both."

Switchgear can also be used in residential and commercial settings to transfer power when a generator is part of the system, such as an automated transfer switch (ATS).

3.2.7 Optimizers

Optimizers are DC-to-DC converters which improve PV module performance via maximum power point tracking. This smooths and regulates the voltage leaving the module. Each optimizer reports module performance data to a smart inverter, which typically is connected to a wireless or cellular network. The inverter reports the data to a software application. A PV module with an optimizer is called a *smart module*.

Most optimizers adhere to the PV Hazard Control System (PVHCS) guidelines and fulfill *NEC®* requirements for rapid shutdown in the event of an emergency. This isolates PV modules from the PV system.

Some optimizers are a 2:1 installation, meaning that two modules can be connected to a single optimizer near the modules. For example, if you have 400 PV modules and need 200 optimizers at roughly $100 each, the cost of optimizers can quickly exceed the cost of an inverter.

3.0.0 Section Review Questions

1. The mechanical BOS components support the primary components based on the engineer's specifications outlined in the _____.
   a. racking hardware
   b. site assessment
   c. solar design
   d. PV module's nameplate

2. Which of the following is the largest gauge wire?
   a. 2 AWG
   b. 6 AWG
   c. 12 AWG
   d. 16 AWG

3. Which of the following is an electrical BOS component?
   a. A solar battery
   b. A PV module
   c. An inverter
   d. A combiner box

4.0.0 Safety Considerations in PV Systems

Objective

Understand safety considerations in PV systems.

a. Describe PPE for fall protection.
b. Understand the potential electrical hazards in working with solar PV systems.

Performance Tasks

This is a knowledge-based module. There are no Performance Tasks.

PV system installation often involves working in extreme heat on slanted, slippery roof surfaces 10 or more feet above the ground. Always wear sunscreen, sunglasses, and non-slip footwear.

Also, extend the top of the ladder 3 feet (36 inches) beyond the roof surface.

Maintain control over all tools, equipment, and system components, so they do not fall or slide off the roof, possibly injuring workers below. Lift from the legs, not the back (*Figure 48*). Wear a hard hat Type E for electrical hazards.

CAUTION

The proper angle for setting up a straight ladder is to place its base a quarter of the working length of the ladder from the wall or other vertical surface. The Occupational Safety and Health Administration (OSHA) calls this the 4:1 ratio or 4 to 1 rule.

WARNING!

Avoid working in wet conditions. Never work where you can see lightning or hear thunder: "If you hear it, fear it; if you see it, flee it." PV modules can act as sails, so be careful when handling them in windy conditions.

Never look directly into the Sun. Avoid excessive sunlight exposure and stay hydrated. Failure to drink enough water can result in heat exhaustion or a heat-related illness that could lead to heat stroke and possible death. Also, avoid using caffeinated drinks or alcohol before or during work.

In addition to working at elevation, you will be handling large, sharp, unwieldy PV modules that have the capacity to become energized the moment light strikes them (*Figure 49*). Keep PV modules covered per the manufacturer's instructions, so they remain de-energized. Double glove by wearing rubber-insulated gloves to protect from electrocution and leather gloves on top of those to prevent punctures. Seek assistance when lifting to avoid back strain.

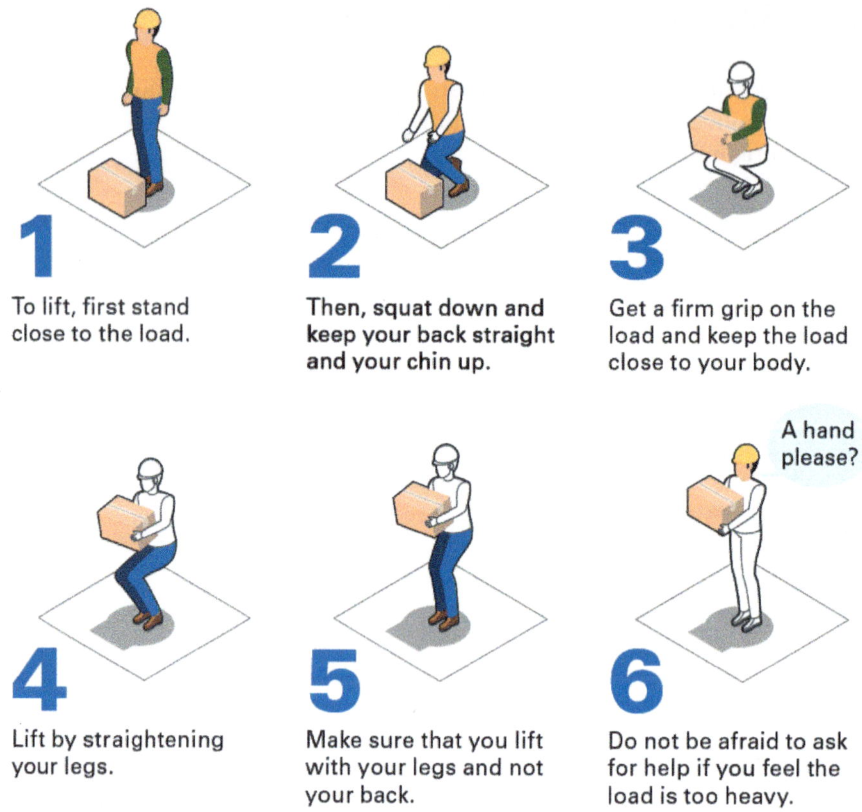

Figure 48 Lift from the legs.

WARNING!

Plugging an individual PV module's positive and negative MC4 connector together, with itself, may cause an arc, increase risk of shock, and damage the connector without a visible indication of the damage.

Figure 49 Installing PV modules.
Source: Chatchawal Phumkaew/Shutterstock

PV modules may be hoisted onto a scaffold or moved using a material handling system. Scaffolds may only be erected and inspected by qualified individuals. All scaffold workers must be trained in the safe use of this equipment. Follow all safety procedures when using mechanical moving equipment. Never operate any equipment unless you are fully qualified.

Working with PV systems may require working in hot attic areas. Provide ventilation where possible and avoid contact with insulation. Identify exit routes before working in tight areas. Exercise caution when working in attic spaces to avoid falling through the ceiling. Install wood across ceiling joists to provide secure footing.

WARNING!

Only qualified individuals may install PV system arrays. Wear all the required personal protective equipment for the task you are performing. Apply sunscreen and stay hydrated when working in hot temperatures. Always use fall protection on work surfaces six feet or more above the ground. Do not step on or rest objects on PV modules. Be aware of all electrical hazards.

4.1.0 Fall Protection

Falls are the most common safety hazard when installing PV systems on an elevated structure. Fall protection is an essential component of roofing safety.

OSHA 1926.501 requires employers to implement fall-protection systems any time an employee is exposed to a walking or working surface at elevations of 6 feet or greater. A certified person with demonstrated competency in the specified fall-protection system installs the safety equipment at the jobsite. Some common fall-protection systems include the following:

- Personal fall arrest system (PFAS) with anchorage points at the site
- Guardrails
- Guarding screens
- Warning lines
- Safety monitoring
- Safety nets

Various kinds of fall protection are required depending on the slope of the roof. The NCCER Roofing Program defines low- and steep-slope roofs as follows:

- *Low-slope* — less than 4:12 slope
- *Steep-slope* — 4:12 slope or more

Depending on the project, low-slope roof systems may require a range of fall-protection methods. Some projects benefit from a combination of available safety options; for example, a low-slope roofing safety plan may combine guardrails with a warning-line system.

Warning-line systems are usually required to be placed around the entire low-slope work zone, whether that area encompasses the whole roof or just a section of it. If the roof is 50′ (15.2 m) wide or less, a safety-monitoring system may be used alone in place of warning lines.

If a low-slope roof requires a worker's attention outside of the warning-line zone, it is crucial that an appointed safety monitor supervises that worker until he or she returns to the area secured by warning lines.

Steep-slope construction has fewer safety options than low-slope construction. Steep-slope roofing reduces available fall-protection systems to just include PFAS, safety-net systems, and guardrail systems with toeboards. Temporary platforms may need to be constructed to provide additional support and fall protection.

If a guardrail system is used for fall protection in steep-slope roofing, OSHA mandates that additional toeboards must be erected along the perimeters of a

Fall Protection Saves Lives

In 2018, the Occupational Safety and Health Administration (OSHA) reported that deaths caused by falls accounted for 33.5% (roughly one-third) of construction-related fatalities. Nearly 90% of these fatal falls occurred when there was no fall-protection system in place.

working surface to prevent tools or debris from sliding down the roof and creating a fall hazard to those walking below the working area.

OSHA standards require PFAS to control falls for any working surface six feet or more above the ground. If you are forced to attach below the D-ring on your harness, you will fall that distance plus the six-foot length of a standard lanyard. This could result in a fall of up to 12 feet. A PFAS is not designed to meet those forces and will transfer much higher impacts when it arrests the fall. To combat this, the American National Standards Institute (ANSI) has changed the equipment manufacturing standards for fall protection. The standard is commonly referred to as the 12-foot lanyard. The lanyard is still only six feet long, but the allowable deployment (extendable length) of the shock absorber has been increased. This lessens the force transferred to the body through the harness, reducing the likelihood of injury.

All personal fall arrest systems must have the ABCs of Fall Arrest Systems (*Figure 50*), which are described as follows:

- *Anchorage/anchorage connector* — Anchorage connectors secure a worker's lanyard or lifeline to the anchorage point. While planning a fall arrest system, the anchorage connector must be carefully selected and placed. It must be easily accessible, safely located, and capable of supporting at least 5,000 pounds (2,250 kg) per employee attached in case of a fall.
- *Body wear* — A full-body harness evenly distributes potential fall force over the user's upper thighs, chest, and shoulders. This harness also provides a point of connection, such as a D-ring that connects the user to his or her linked device. A D-ring can always be found integrated into the back of a full-body harness; however, harnesses may also have additional D-rings located around the shoulder, hip, or sternum areas.

NOTE

In the event a safe anchorage point is not available, specialized equipment may be brought to the jobsite to temporarily create an engineered anchorage point. Following *OSHA 1910.140(d)*, the load capacity requirement for this specialized equipment is reduced from 5,000 pounds to 3,600 pounds.

Figure 50 ABCs of PFASs.

- *Connecting devices* — Connecting devices, such as a lanyard or lifeline, connect a user's body harness to a designated anchorage point. If a fall occurs, the connecting device contains a braking mechanism that typically activates within three feet of a drop to reduce movement impact and prevent swinging falls.

Before using a PFAS, you must examine the space below any potential fall point. Make sure it is clear of any hazards or obstructions. When assessing the required free length of a PFAS, you must account for the deployment of the shock absorber. In addition, PFAS manufacturers require up to a three-foot safety factor. Taking this safety factor into account, *Table 13* indicates the minimum distance required between any hazard/obstruction and the worker or attachment point.

A typical single-story building may not provide enough height to use standard lanyards. Retractable lanyards that engage and stop falls in less than two feet are usually the best choice for fall protection when used below 20 feet or on any type of lift equipment.

The weakest link in the PFAS is the locking mechanism (gate) on snap hooks. OSHA requires the gate to be rated for 350 pounds when the system is rated at 5,000 pounds. In response, ANSI issued a standard requiring the gate strength to be a minimum of 3,600 pounds.

NOTE

Ladders, scaffolds, stairs, and similar types of equipment have other fall protection standards. Make sure you know and follow your company safety policies when using this equipment.

TABLE 13 Required Free Lengths for Various Lanyards

Lanyard Type	Lanyard Length	Shock Absorber Length	Average Worker Height	Safety Factor	Required Free Length
Standard 6' lanyard with 3.5' shock absorber	6'	3.5'	6'	3'	18.5'
ANSI 6' lanyard with 4' shock absorber	6'	4'	6'	3'	19'
ANSI 12' lanyard with 5' shock absorber	6' (12' freefall)	5'	N/A (accounted for by lanyard)	3'	20'

4.2.0 Electrical Hazards

Before working on any electrical equipment, the equipment and all inputs and outputs must be placed into an electrically safe working condition. This means that all equipment must be de-energized, any stored electrical energy must be discharged, and the equipment must be locked out and tagged (*Figure 51*) before any work can be performed. A licensed electrician performs a "Live-Dead-Live" safety test to confirm that electricity is not flowing through the equipment.

4.2.1 Electrical Shock

Remember that PV systems have multiple power sources. These include utility power, PV system power, and batteries, if used. In actual use, a PV module may experience conditions that result in more current and/or voltage than listed on the PV module's nameplate.

A PV module generates electricity when exposed to sunlight. Keep the PV module covered until ready for installation and use insulated tools. When the PV module is removed from the shipping container, avoid touching the terminals together when handling.

Some of the most common electrical hazards in PV systems include cracked or broken PV modules, faulty connections, broken ground wires, and undersized wiring. Be aware of all potential hazards and exercise caution during installation and troubleshooting.

Where it is not possible to eliminate a hazard, an energized electrical work permit must be completed, and workers must be protected by appropriate safety procedures and personal protective equipment. Use insulated tools.

Figure 51 Lockout tagout.
Source: ArtboyAnimation/Getty Images

"Live-Dead-Live" Safety Test

A "Live-Dead-Live" safety test is used to verify the absence of voltage in a system before performing work. It is mandated by OSHA when voltages are above 600 volts and requires specific personal protective equipment (PPE). It is also required by the NFPA® 70E in *Section 110.4(A)(5), Operation Verification*, as well as *Section 120.1(5), Verification of an Electrically Safe Work Condition*.

Only a licensed electrician performs this safety test.

Backfeed: The flow of electric power in the direction reverse to that of the generally understood or typical flow of power.

Backfeed

Electrical **backfeed** occurs when electricity flows in the reverse direction of what is expected in an electrical system in a given moment. An engine-driven power generator can cause backfeed in a home when plugged into an outlet that isn't intended for that use. During a power outage, when utility workers are working on the broken power lines, a solar PV system backfeeds electricity that it has generated into the power grid. Safety shutoffs prevent this from happening.

NOTE

For more information on electrical safety, reference NCCER's *Managing Electrical Hazards*. This program describes how to assess and eliminate shock, arc blast, and arc flash hazards using the practical and safe working requirements detailed in *NFPA 70E: Standard for Electrical Safety in the Workplace®*. Also, reference OSHA's Electric Power Generation, Transmission and Distribution standard, *29 CFR 1910.269*, for safe work practices and worker training requirements.

Multimeter Safety

A multimeter is used to determine if a circuit has electricity. It can be set to measure voltage or amperage. Before using any multimeter, ensure that it is rated for the given application. PV installations are high voltage systems and require the use of Category III/IV meters. Inspect the multimeter and leads for damage before use. Never use a damaged meter. Wear safety glasses when using a multimeter.

4.2.2 Arc Flash

Arc flash is the light and heat from electricity as it leaves its intended circuit, flowing through an air gap to another conductor or ground. It occurs when the voltage potential exceeds the initial conductor or there is a broken wire. This happens mainly on the DC side of a solar PV system: DC wiring, battery storage, and electric vehicles and their charging infrastructure.

Arc flash doesn't typically occur on the AC side of the solar PV system because electrical codes and the design of AC components have safety built in. Also, 120V does not have the capacity to cause an arc flash. However, if a building has 480V service, then it is possible.

Most faults can be categorized as follows:

- Line to ground
- Line to line
- Module to module
- String to string

NOTE

String fuses protect against backfeed and array faults.

Figure 52 shows an example warning label, common DC arc flash locations, or faults, and PPE.

4.2.3 Battery Hazards

Flooded lead-acid (FLA) batteries release hydrogen when they are overcharged. This gas can build up to unsafe levels if batteries are poorly vented. The pressure rises within the cells until a spark or short circuit ignites the gas and it explodes. Battery explosions can cause severe injury or death. Replace any battery that feels unusually warm, appears swollen, or leaks.

When handling batteries, wear approved, chemical-rated safety equipment, including eye and face protection, gloves, and aprons.

Solar PV systems sometimes require that batteries be connected in a series. This series causes increased voltage between the end terminals that may cause an electric shock if touched. Always be sure to handle batteries with proper care.

Most batteries contain toxic elements that are harmful to the environment and may pose a fire risk. Contact your local waste-management office to properly dispose of batteries and, in some cases, recycle.

NOTE

For more information concerning energy storage system (ESS) handling and hazards, consult *NEC Article 480* and *NEC Article 706*. These articles address items regarding energy storage systems, such as equipment, circuit sizing, charge control, and battery interconnection. *NEC Article 625, Electric Vehicle Power Transfer System*, may apply.

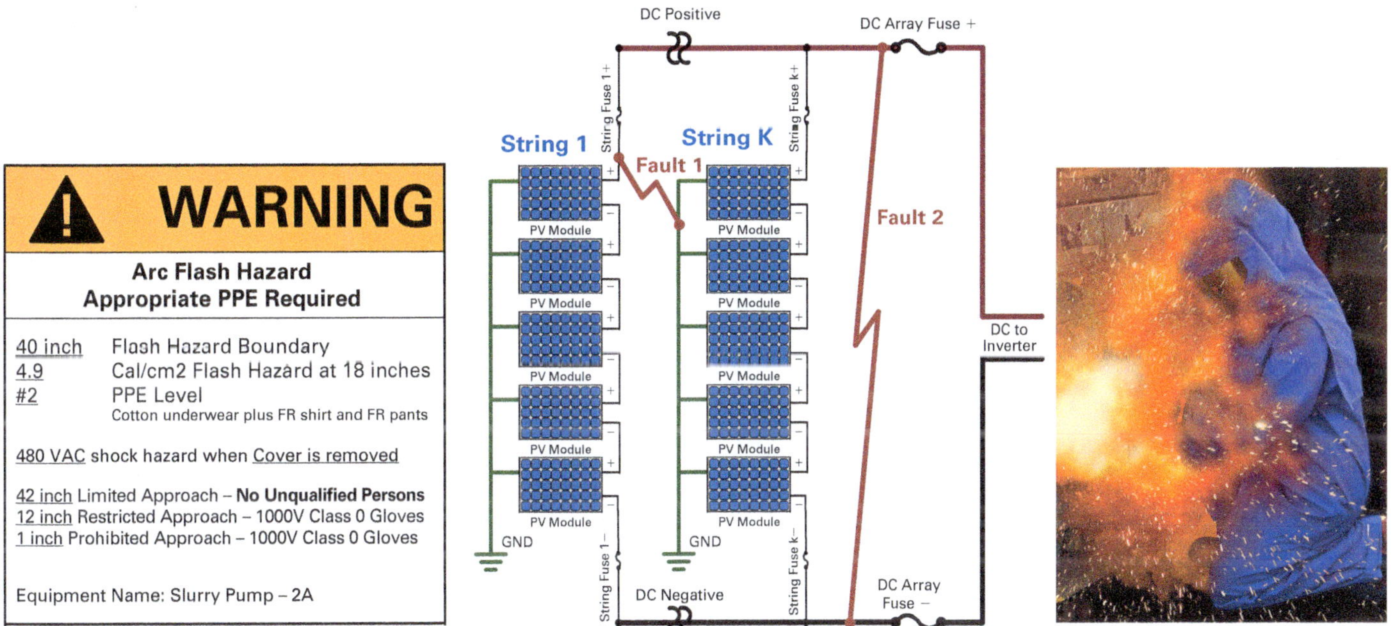

(A) (B) (C)

Figure 52 Reading arc flash warning labels and possible DC arc locations.
Source: (A) OSHA; (B) NREL

Battery Temperature

Batteries rise in temperature prior to failure. Especially with lithium-ion batteries, if the internal cells become damaged because of the heat, there is an increased risk of an explosion.

A battery temperature sensor can be used with certain charge controllers to monitor the batteries and shut down the system at excessive temperatures. As shown in *Figure 53*, infrared or thermal imaging helps troubleshoot battery problems. The batteries in the top row aren't charging; note the cooler color in the thermal imagery.

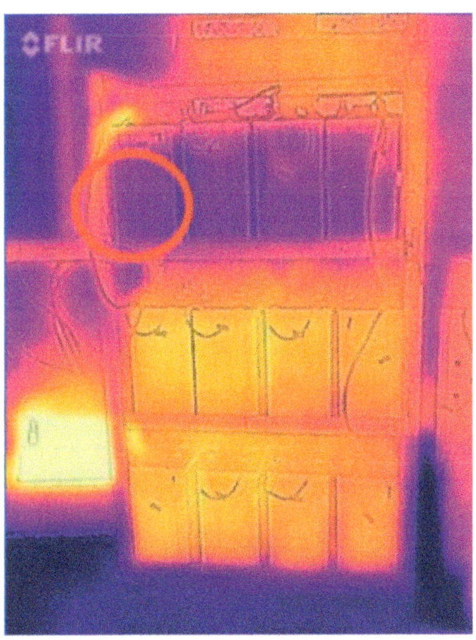

Figure 53 Troubleshooting with thermal imaging.
Source: Owen Roberts, National Renewable Energy Lab (NREL) https://www.nrel.gov/docs/fy19osti/73822.pdf

WARNING!

Placing a wire across the terminals of a battery is called *shorting the battery*. Some batteries will explode when shorted. Never short a battery.

4.0.0 Section Review Questions

1. OSHA requires employers to implement fall-protection systems any time an employee is working at elevations of _____.
   a. 3 feet or greater
   b. 6 feet or greater
   c. 8 feet or greater
   d. 10 feet or greater

2. What test does a licensed electrician perform to confirm that electricity is not flowing through the electrical equipment?
   a. "Dead" safety test
   b. "Live" safety test
   c. "Live-Dead-Live" safety test
   d. "Shutdown" safety test

3. What is one symptom used in diagnosing battery failure?
   a. Decrease in battery temperature
   b. Increase in battery temperature
   c. "Live-Dead-Live" safety test
   d. "Shutdown" safety test

5.0.0 PV System Project Life Cycle

Performance Tasks

This is a knowledge-based module. There are no Performance Tasks.

Objective

Describe the project life cycle of a solar PV installation.

a. Understand how a utility bill helps establish customer solar energy goals.
b. Identify key steps in performing a solar site assessment.
c. Describe key factors involved in decision making for a solar system design.
d. Describe the key steps in system installation.
e. Explain what to look for in a system maintenance visual inspection.

A solar photovoltaic (PV) system installation job has five main phases in the project life cycle (*Table 14*).

Realize that in your first job, you may be assigned to one dedicated task within one specific step in the process. The entire cycle is covered here to expose you to a few career paths available in solar PV and to help in your decision-making. Which career direction is right for you?

TABLE 14 Project Life Cycle

Project Life Cycle Phase	Main Objective
Sales	Establish the customer's energy efficiency and solar energy goals, which drive all decision making.
Site assessment	Assess the solar access and available physical space of a jobsite to propose potential solar PV installation sites.
System design	Based on the customer's energy goals and the site assessment report, create a solar system design, which includes sizing the system, recommending the products to be used, and developing electrical schematics for the design.
Installation	Install the solar PV system according to the system design, from creating foundations and assembling parts to pulling non-live wires into place.
Maintenance and troubleshooting	Service an existing solar PV system.

5.1.0 Setting Solar Energy Goals

During a site assessment, a salesperson or site assessor interviews the customer to accomplish the following:

- Educate and set expectations regarding energy efficiency measures and solar PV energy.
- Establish the customer's solar PV energy goals, which drive all decision making for the project.

Customer service is key. When interacting with customers, be sure to project and maintain professionalism. Adapt to your audience when speaking: A residential homeowner installing rooftop solar will have different customer needs than a landowner looking to install a utility-scale system. Always be patient and attentive. Ask questions about any gray areas in communication.

Energy Efficiency

Energy efficiency measures are introduced to decrease the building's energy consumption. This is an optimization step prior to the sizing and installation of the solar PV system.

Energy Goal Setting

During the interview, discuss the following key factors:

- Use utility bills to assess annual energy consumption and establish PV system sizing goals.
- Discuss net metering and what percentage of grid power replacement the customer is seeking.
- Ask if the customer has an electric vehicle or requires energy storage.
    - If energy storage is requested, establish a critical loads list to help size the battery bank.

NOTE

There are many optimization strategies for aligning the solar PV arrays with the Sun to meet the customer's solar PV energy goals. For more details, consult NCCER Module 57102, *Site Assessment*.

Roofing

For roof-mounted installations, discuss the condition and age of the roof, matching the roof's life expectancy with that of the solar array's life expectancy. If the roof is more than 10 years old, then it may be cost effective for the owner to replace the roof before installing solar PV. Also, review the roofing manufacturer's warranty to understand possible restrictions and remain compliant.

Financing

Some solar design software solutions include financial loan options with the proposal documentation. This helps customers make informed decisions about their investment. Rebates, tax incentives, and net-metering options may also assist in the financial decision-making process, and typically require the use of the charted solar window for estimating potential system energy output.

5.1.1 Energy Consumption

Use the customer's utility bills in the following steps to determine the building's energy usage details.

Step 1 Add up the monthly energy consumption to get the annual energy consumption.

Step 2 Determine average monthly and daily energy consumption from the annual value.

Step 3 Calculate the number of PV modules required to meet that demand.

If bills aren't available, most utility companies offer their customers free energy audits.

The building energy usage information is in kilowatts and kilowatt-hours. Appliances and lighting, also known as *electrical loads*, use power to do work, and that power is measured in watts.

- 1,000 watts (W) equals 1 kilowatt (kW).
- Kilowatt-hours (kWh) is the amount of electricity used over time.
- One kWh equals 1,000 watts per hour of power used.

The sample utility bill (*Figure 54*) shows an electric consumption of 903kWh for the billing month of May. The bar graph shows estimated annual energy consumption, and July looks to be the most demanding month of the year at roughly 1,523kWh. Again, you'd ask the customer for 12 months of bills and add up the monthly usage to get the annual energy consumption.

According to the US Energy Information Administration, in 2020, the average annual electricity consumption for a US residential utility customer was 10,715 kilowatt-hours (kWh), an average of about 893kWh per month or 29.36kWh per day.

Due to cost, some grid-tied customers may not want a solar system that covers 100% of their energy demands, so adjust to reflect the customer's solar PV energy goals. For example, if customers want to cover half of their energy needs, then multiply their energy consumption by 50% before sizing the PV system.

Sizing the Battery Bank

Sizing the battery bank requires a list of critical loads (*Table 15*), such as appliances and lighting, that need to be powered by the batteries when the solar PV system isn't generating electricity.

Account: 2000-1234-5678 Bill Date: 05/26/2017 Page 2 of 2

Electric Service Details Example

Meter Number	METER READING DATES Previous	Present	Meter Read Type	Days of Service	METER READINGS Previous	Present	Meter Multiplier	Electric Consumption	Peak Demand	Next Meter Read on or About
E112233	4/18/2017	5/19/2017	ACTUAL	31	73670	74573	1.00000	903 kWh	N/A	6/19/2017

Description	Usage	Rate	Amount
ELECTRIC CUSTOMER CHARGE			$ 14.25
ENERGY USE, TIER 1 (1 – 850 kWh)	850	$ 0.0430	$ 36.55
ENERGY USE, TIER 2 (OVER 850 kWh)	53	$ 0.0640	$ 3.39
ELECTRIC FUEL ADJUSTMENT	903	$ 0.0700	$ 63.21
FLORIDA GROSS RECEIPTS TAX			$ 3.01
GAINESVILLE ELEC UTIL TAX			$ 5.72
Total for Electric Services			**$ 126.13**

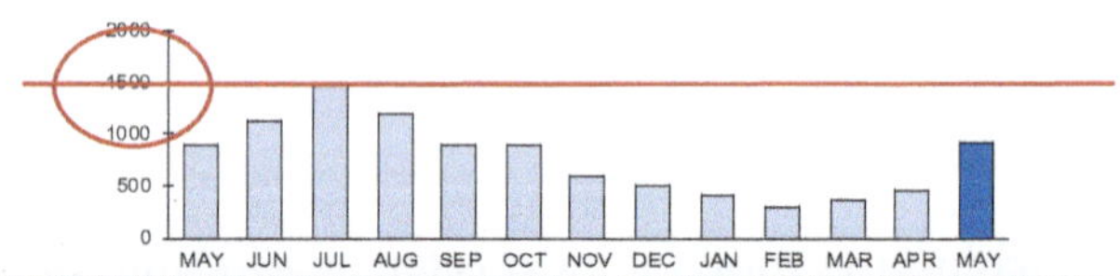

Gas Service Details Example

Meter Number	METER READING DATES Previous	Present	Meter Read Type	Days of Service	METER READINGS Previous	Present	Meter Multiplier	BTU Factor	Gas Consumption	Next Meter Read on or About
G223311	4/18/2017	5/19/2017	ACTUAL	31	3204	3221	1.01700	1.020	18 THM	6/19/2017

Description	Usage	Rate	Amount
NATURAL GAS CUSTOMER CHARGE			$ 9.75
NATURAL GAS USE	18	$ 0.6300	$ 11.34
MANUFACTURED GAS PLANT REC	18	$ 0.0556	$ 1.00
PURCHASED GAS ADJUSTMENT	18	$ 0.2300	$ 4.14
FLORIDA GROSS RECEIPTS TAX			$ 0.85
GAINESVILLE GAS UTIL TAX			$ 2.29
Total for Gas Services			**$ 29.37**

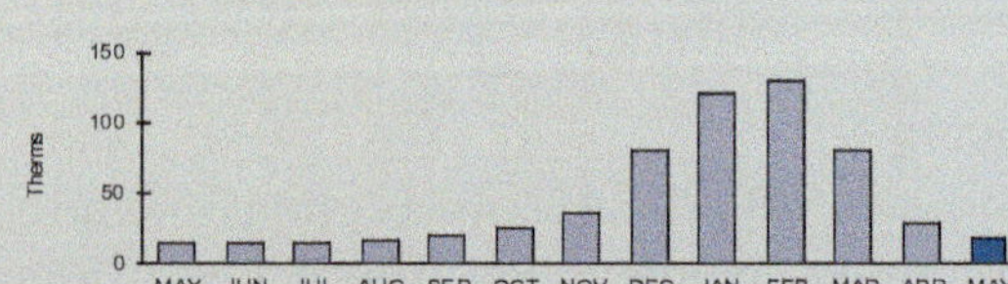

Water and Wastewater Service Details Example

Meter Number	METER READING DATES Previous	Present	Meter Read Type	Days of Service	METER READINGS Previous	Present	Meter Multiplier	Water Consumption	Wastewater Consumption	Next Meter Read on or About
W331122	4/18/2017	5/19/2017	ACTUAL	31	1101	1113	1.00000	12 kGals	8 kGals	6/19/2017

Description	Usage	Rate	Amount
WATER CUSTOMER CHARGE			$ 9.45
WATER USE, TIER 1 (1 – 4 kGals)	4	$ 2.45	$ 9.80
WATER USE, TIER 2 (5 – 16 kGals)	8	$ 3.75	$ 30.00
WATER USE, TIER 3 (OVER 16 kGals)	0	$ 6.00	$
GAINESVILLE WATER UTIL TAX			$ 4.93
WASTEWATER CUSTOMER CHARGE			$ 9.10
WASTEWATER BILLING	8	$ 6.30	$ 50.40
Winter max is 8 kGals			
Total for Water Service Details			**$ 113.68**

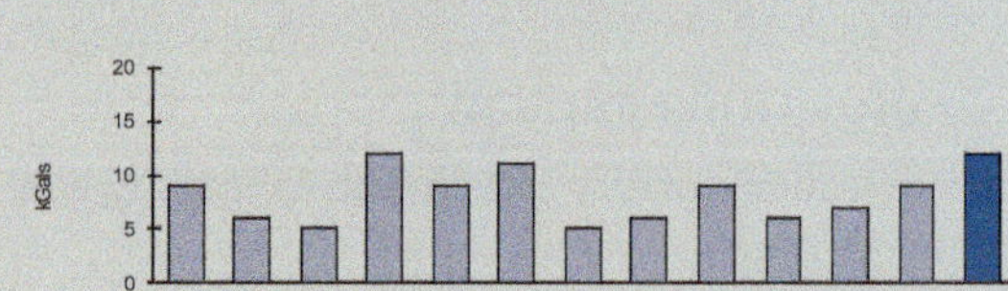

Figure 54 Reading a utility bill.

Source: Based on https://www.gru.com/LinkClick.aspx?fileticket=kmvFZg3GbW8%3d&portalid=0

TABLE 15 Example Critical Loads List

Appliance	Watts	Estimated Usage Hours in One Day	Watt-Hours
Refrigerator	107	12	1,284
Space heater	1320	8	10,560
Coffee maker	1000	0.25	250
String inverter	15	24	360
Total demand			12,454 watt-hours per day **12.5kWh/day**

Think About It

Add Two Light Bulbs

Two light bulbs were left off the list in *Table 15*. Look up the wattage of a light bulb and estimate how many hours it would be used during the day, then update the critical loads list. Note that the inverter was included in the list.

A critical design month must be chosen. In cold regions, this may be a winter month, because this time of year has less daylight. For each load, estimate the daily hours of use during that month and multiply this value by the equipment's nameplate wattage to determine the total demand in watt-hours (Wh). If a nameplate is unavailable, make load estimations with the Department of Energy's (DOE) online Appliance Energy Calculator. The more loads in the list, the more expensive the battery bank will be to support the list.

NOTE

Appliance wattages can usually be determined from the owner's manual or equipment nameplate. If an appliance is rated in amps, use Watt's law. Multiply the amperage by the operating voltage to determine the value in watts.

5.2.0 Site Assessment

A site assessor reviews the customer's solar PV energy goals and assesses what is possible with the given space at the jobsite and the access point to the power grid.

Before visiting the jobsite, use online tools to research environmental factors for that location (e.g., solar irradiance, wind speeds, snow loads), look up specific city/county setback requirements, and find out what utility company the building most likely uses. Online tools can help you with site assessment tasks, such as uncovering solar data for a specific location, measuring solar irradiation, and assessing the total potential solar energy of a site. Draft a preliminary site survey using Google Earth aerial photos and use this to take detailed notes while at the site.

At the jobsite, assess the site for optimal solar access. In the Northern Hemisphere, using a compass, find true south to establish basic array orientation with the Sun and choose southern facing roofing surfaces, if applicable.

NOTE

This introductory module is designed to give you a basic overview of solar installation. For more details about site assessment, consult NCCER Module 57102, *Site Assessment*.

WARNING!

When performing a site assessment, always wear sunlight protection, protective sunglasses, and shoes with non-slip soles. Use the appropriate fall protection and ensure that straight ladders extend beyond the roof surface by 36". To avoid permanent eye damage, never look directly into the Sun.

Using the site assessment checklist and your preliminary site survey, document details about the site and take video and photos of your findings. Assess the site for sufficient physical space for the solar array(s) and a laydown area for organizing equipment and materials the day of installation. Think about where the BOS components could be supported and grouped together. For a ground-mounted installation, look for the closest point of interconnection. Consider the effects of severe weather at the location and the layout of the PV modules. Plan for soil studies and review of water table levels to assist in making foundational and structural decisions.

In the case of a roof installation, document the type, size, age, and condition of the roof, including the integrity of support members.

CAUTION

For your safety, inspect the interior, then the exterior of buildings for potential problems with the mounting surface.

Shade Analysis

Perform shade analysis, measure the height of obstructions, and document findings. Take photos and sketch on a sun path chart. Recall that shadows are longest on the shortest day of the year.

Think About It
Perspective

A site assessor may begin PV array sizing from the perspective of how much space is available to mount PV modules, while a system designer may begin with the customer's PV energy goals.

The information you gather at the jobsite is entered into solar photovoltaic system design software. The more accurate your information, the better. Shade analysis can also be performed within solar design software if the height of obstructions is provided.

How Much Space Is Enough?

Whether ground mounted or roof mounted, is the proposed space sufficient for the array and desired energy output? Also, how does the planned energy output compare to the building's annual energy consumption?

There are a few ways to look at spatial relationships of the observed physical space and the PV module's size. *Figure 55* shows an example at the NCCER headquarters. NCCER is considering adding a new solar PV array into a vacant strip of their land, adjacent to the company parking lot. The land has a length of 200 feet by a width of 50 feet, which is 10,000 square feet. This is roughly the size of two basketball courts placed end to end. It is an open field with no shade issues. The solar PV professional is recommending a 400W PV module that is 6.5 feet by 3.5 feet in size, which is 22.75 square feet.

There are three ways to estimate the number of PV modules that would fit in the field for a ground-mounted installation:

- Square footage
- Landscape orientation
- Portrait orientation

As you work through the estimations, as shown in the figure, compare the total number of modules estimated by each calculation. The square footage estimation doesn't maintain the physical shape of the PV modules or the field's boundaries. This method overestimates the number of PV modules that can fit in the space.

The landscape and portrait orientation calculations respect the actual dimensions of the PV module and the field's boundaries, so these estimates are closer to what can actually be installed. What is not included is any required maintenance access pathways or fire code guidelines, which might eliminate modules from the potential layout.

Fire Access Example

Surfaces longer than 150 feet in any direction require the arrays to be grouped into 150 foot by 150 foot sections. This means every 150 feet there is a walkway of either 48 inches or 96 inches in width, depending on the available roof openings, such as ventilation ducts or skylights. If available, fire crews use roof openings to cut open the roof's structure to fight fires. If they aren't available, then the walkway is built to be 96 inches wide, which provides space for firefighters to cut holes into the roofing surface.

Let's compare these ground-mounted installation estimates to a roof installation. The company roof might be another place to install the PV modules, and it is the same dimensions as that estimated for the ground-mounted installation. However, unlike the open field, this roof has a large obstruction in the center. PV modules cannot be installed on this obstruction, and it will cast shadows on additional roof areas, making those areas unusable. What about the maintenance and emergency access pathways? This is where solar design software is helpful in creating an accurate plan.

Think About It
Space Efficiency

What about a carport installation?

What about estimating the amount of energy the PV system would generate? That depends on where the system is located. NCCER is in Florida with an average of 5.00 peak sun hours per day. Let's compare it with other US locations, like San Antonio, Texas, and Little Falls, Minnesota. As shown in *Figure 56*, a US solar irradiance map is used to look up the peak sun hours for those locations.

A PV system located in San Antonio, TX, would have the highest energy yield of the three locations, with Little Falls, MN, generating the least amount of energy. The average residential 6kW system is provided for comparison purposes.

NOTE

Actual PV module and inverter outputs vary, and this is a general sizing strategy that can be applied to determine whether a roof or field has enough space for the installation of an array. It doesn't take into consideration access pathways, system losses, voltage drop, or inverter efficiency. These topics are covered in NCCER Module 57103, *System Design*.

PV Module Configurations

The physical space that PV modules occupy is influenced by their arrangement. PV modules can be installed in landscape or portrait orientation. They can also have a high-profile (stacked) or low-profile configuration, as shown in *Figure 57*. As you make decisions, especially for a ground-mounted system, remember that the PV modules do have a "sail-like" characteristic. A high-profile configuration may be more affected by wind uplift forces than a low-profile one, making it less suited for higher-rated wind zone locations.

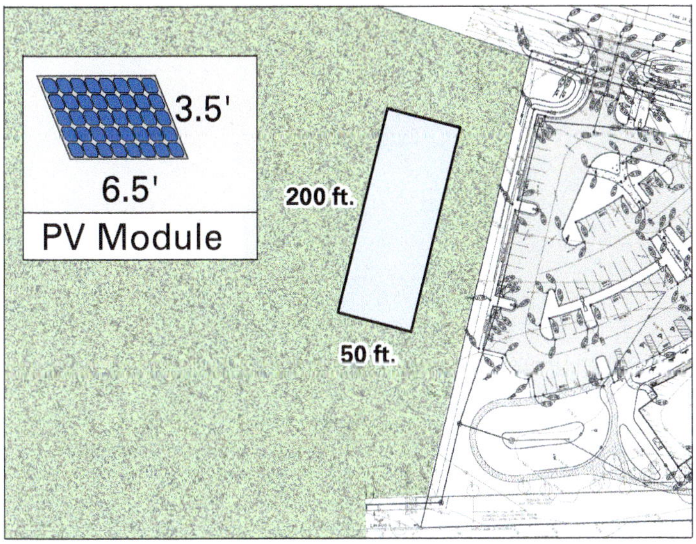

Sqft Estimation

Field: 200 ft. × 50 ft. = 10,000 f^2
Module: 6.5 ft. × 3.5 ft. = 22.75 f^2
10,000 f^2 / 22.75 f^2 = 439.6 module
Round down to 439 modules

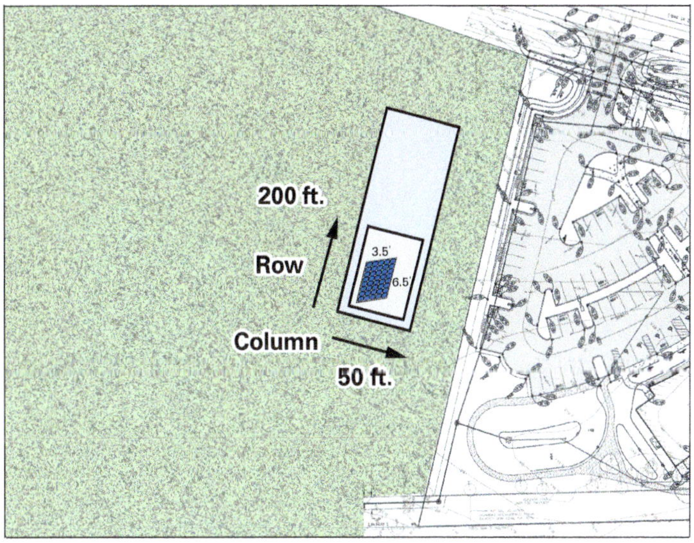

Landscape

200 ft. field length/6.5 ft. module length = 30.76
Round down to 30 modules per row

50 ft. field width/3.5 ft. module width = 14.29
Round down to 14 modules per column

30 × 14 = 420 modules total

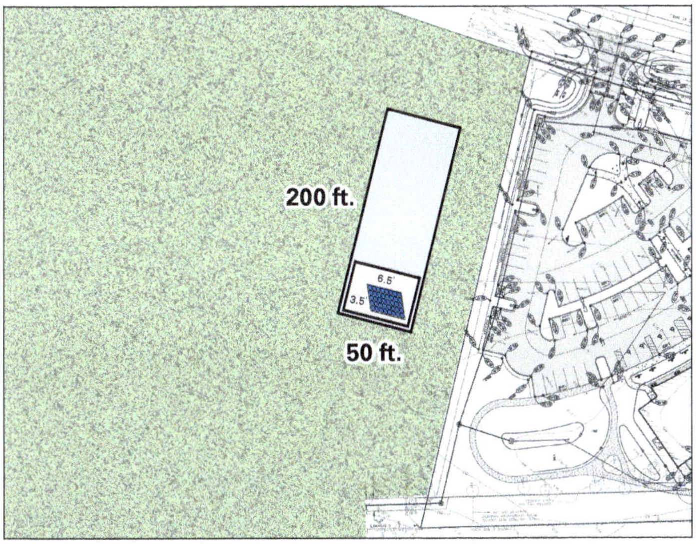

Portrait

200 ft. field length/3.5 ft. module width = 57.14
Round down to 57 modules per row

50 ft. field width/6.5 ft. module width = 7.69
Round down to 7 modules per column

57 × 7 = 399 modules total

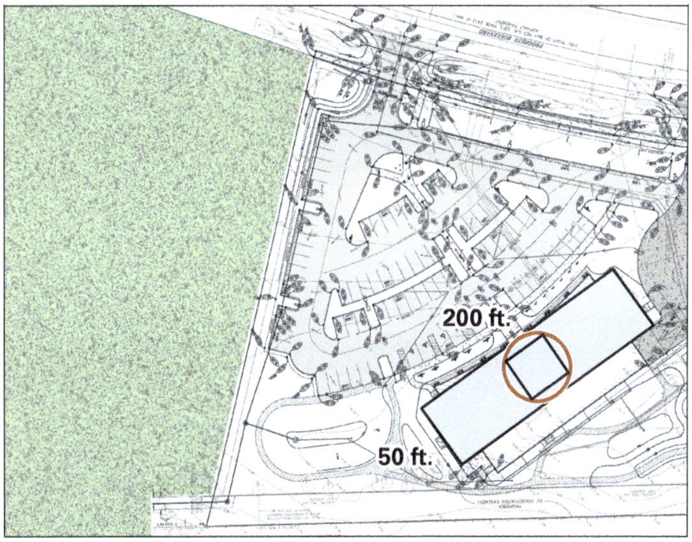

Other Possible Locations?

The company roof might be another place to install the PV module, and it is the same dimensions as that estimated for the ground-mount installations, but wait...there is some kind of protrusion in the middle of the roof, a place where modules couldn't be installed.

Also, what about maintenance and emergency access pathways?

Figure 55 Spatial relationships.

Source: Modeled on results from Google Earth search: https://earth.google.com/web/@29.78018529,-82.47537949,25.37210742a,307.82630944d,35y,70.60040518h,0t,0r

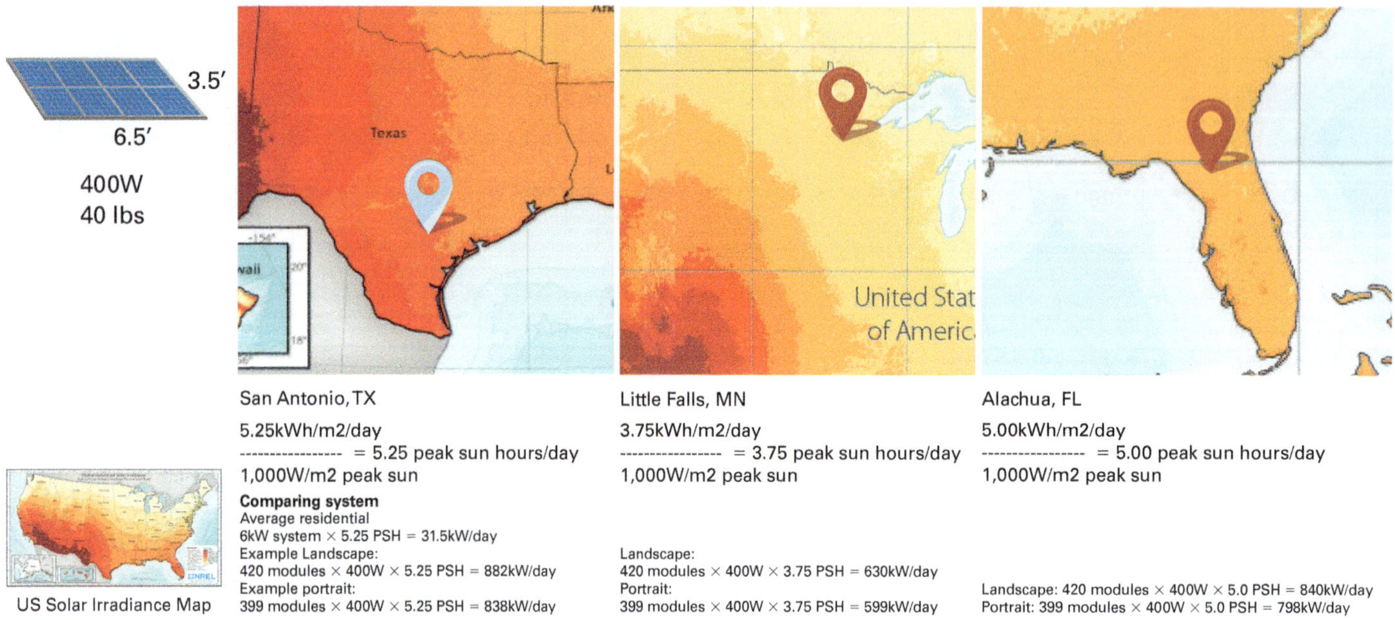

Figure 56 Location comparisons.
Source: NREL

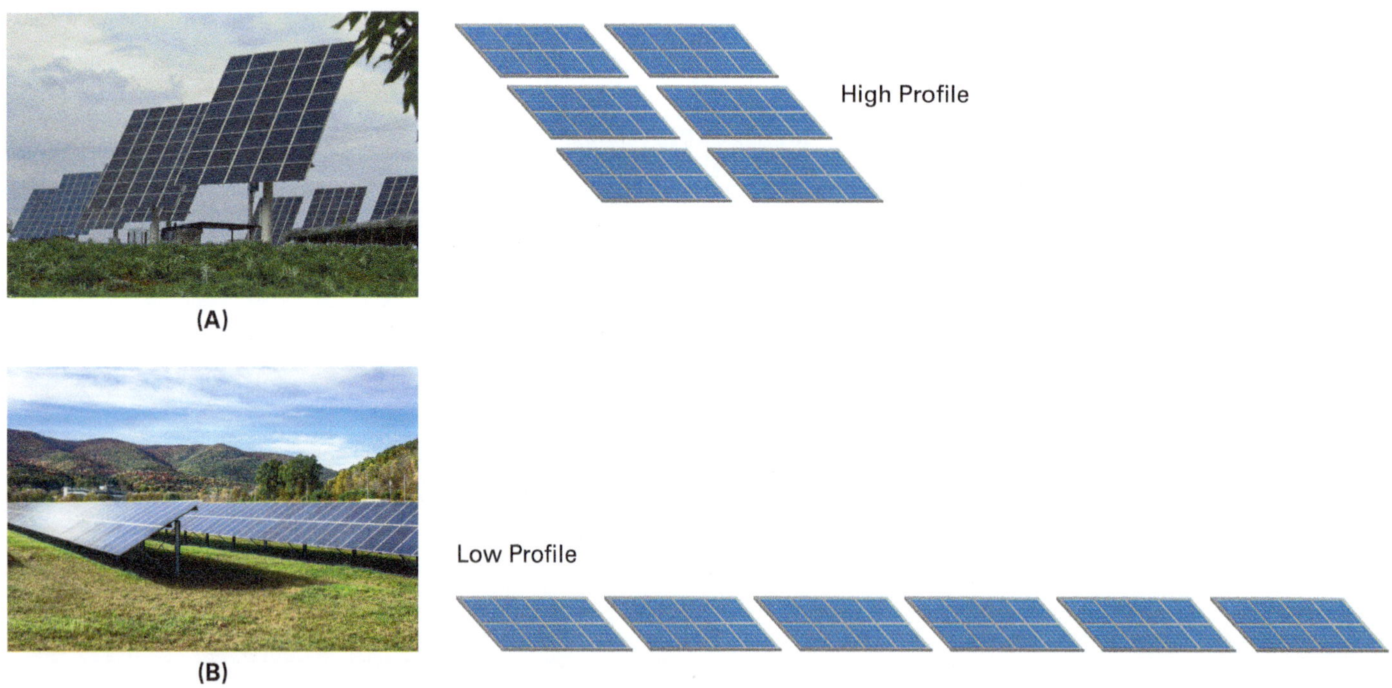

Figure 57 PV module orientation.
Source: (A) Franco Lo/Getty Images; (B) AlbertPego/Getty Images

5.2.1 Roof-Mounted Systems

Check roofing manufacturer requirements to make sure the roofing system is compatible with a solar PV installation and begin relationships with the manufacturer early in the process. Request roofing schematics.

WARNING!

If the roof structure is not sound, inform the building owner. Either a different location must be considered, or the installation must be delayed until the roof is repaired. Never attempt to install a PV array on an unsound structure.

Determine which roofing surfaces have southern exposures that aren't heavily shaded.

5.2.2 PV Components Location

Take photos of the jobsite. Sketch the location of the service in relation to the proposed array. Determine a location for the inverter, DC/AC disconnects, and grounding. Group items together whenever possible to minimize voltage drop within the system.

NOTE

When sketching a rooftop location, consider how and where trays, conduit, and cables traverse and penetrate the roof system. Visualize the maintenance and fire pathways and organize components to minimize walking time.

Energy Storage

For systems that require energy storage, document a suggested charge controller and battery location. Typically, battery banks are located close to the array to minimize voltage drop and need to be protected against temperature extremes. Maintenance-free AGM, gel, and lithium-ion batteries are stored in steel cabinets. Flooded lead-acid (FLA) batteries are stored in vented, high-density polyethylene (HDPE) containers.

For residential systems, electric vehicle owners may opt for their car battery to serve as an energy storage solution for their solar PV system, so placement of the PV components in the garage may be a solution.

Site Assessment Report

Write up all your findings. Compile photos and support documentation into a site assessment report. All details are entered into the solar system design software, so the more detailed, the better.

5.3.0 System Design

PV system design begins with the following information:

- The customer's solar PV goals
- The building's energy consumption, collected during the customer interview
- Jobsite specifics, detailed in the site assessment report

NOTE

This introductory module is designed to give you a basic overview of solar installation. For more details on solar system design, consult NCCER Module 57103, *System Design*.

Review the mounting surface type and condition. For installations in regions that experience severe winds, such as hurricanes or tornados, consult with a licensed structural engineer to calculate any additional anchor points needed to secure the system.

As shown in *Figure 58*, solar system design software (e.g., HelioScope, Aurora) is used for the following tasks:

- Creating PV module layout and electrical schematics
- Calculating number and placement of modules and electrical components
- Establishing a Bill of Materials, used for purchase and inventory of parts

Information from the site assessment report is entered into the software. Use your knowledge of the following factors to make decisions about the array orientation:

- Alignment with the Sun to set the tilt angle and azimuth angle of the arrays
- PV module landscape or portrait orientation

Recall that general recommendations for tilt angle include the following:

- *Maximum output power in winter* — In the Northern Hemisphere during the winter months, the tilt angle should equal local latitude +15°.
- *Maximum output power in summer* — In the Northern Hemisphere during the summer months, the tilt angle should equal local latitude –15°.
- *Seasonal adjustment* — In March and September, the tilt angle equals the local latitude.

These adjustments ensure maximum efficiency at all times.

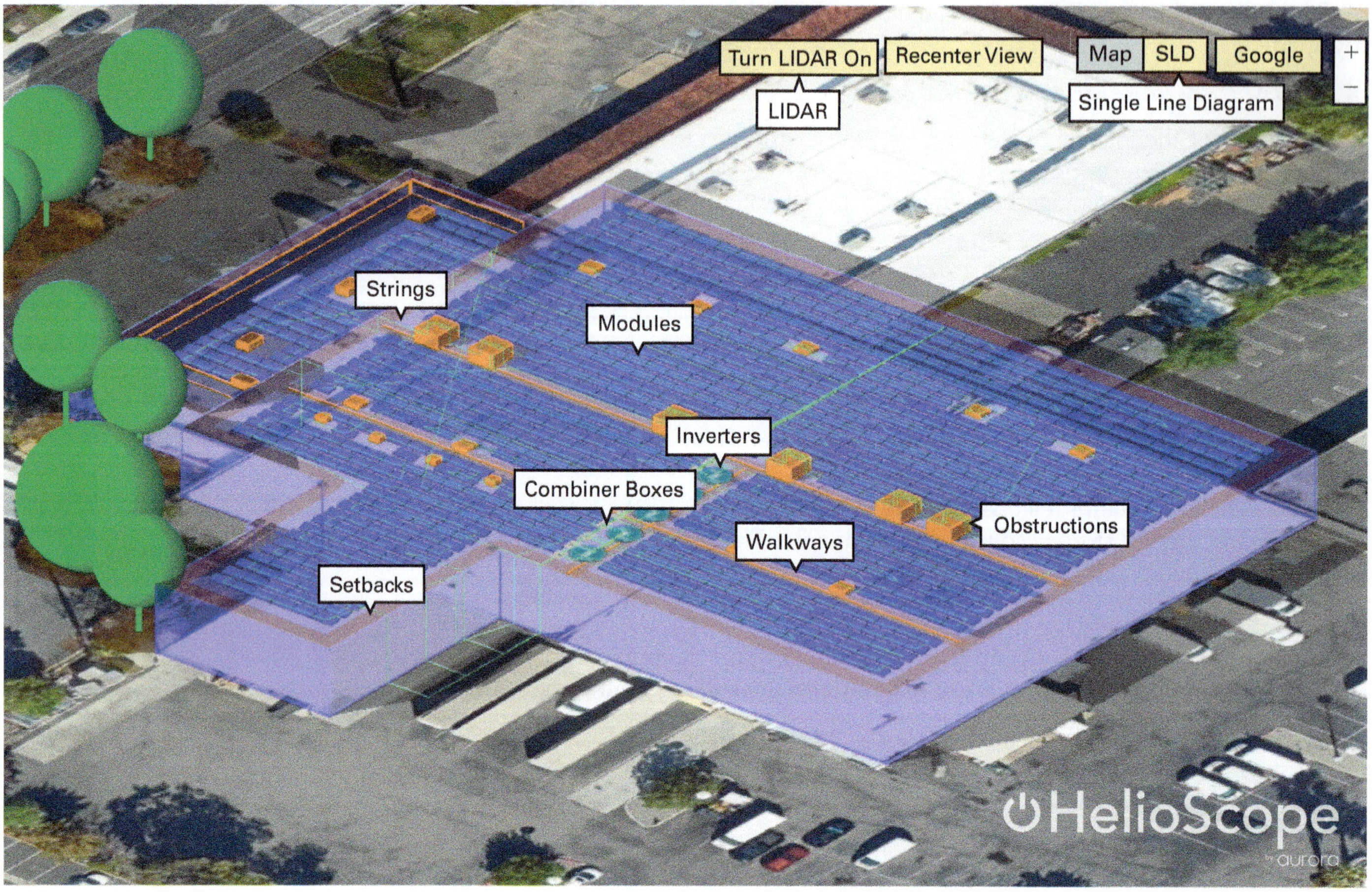

Figure 58 Solar system design software.
Source: Aurora Solar

NOTE

When designing maintenance pathways, keep distances and cable runs between combiner boxes and inverters as short as possible to minimize walking distances during inspections and reduce system losses.

A database of solar PV components is available within the system design software to assist with component selection. Optional energy storage components, such as batteries and charge controllers, are also available. Some products work better together than others, and designing iterations for comparison is part of the job.

The software's tool palette enables the removal of suboptimal PV modules and creation of maintenance and fire pathways to code, among other things.

The design process considers shading. Detailed height measurements of rooftop protrusions (e.g., ventilation ducts, skylights, parapets) taken during the site assessment are entered into system design software to fine-tune the plan. Remember to consider inter-row spacing.

NREL's free *PVWatts*® calculator can also be used to estimate a solar system's energy output for system design.

NOTE

If a roof has slate, clay, or concrete tile, the array installation must be performed by a specialized roofing contractor.

5.3.1 Roofing Considerations

Review roofing manufacturer production specifications and warranties to identify any PV limitations to preserve the warranty.

Roofing structures do not support an unlimited amount of weight. To determine if the building can support the weight of the solar array, roof load capacity should be calculated by a licensed structural engineer. Especially on low-slope commercial buildings, the roof may have become a storage area for heavy, bulky equipment. Additionally, ballasted mounting systems have heavy concrete blocks to factor in.

The tilt angle of PV modules on a steep-slope roof follows the roof's slope (*Table 16*).

Did You Know?

Fine-Tuning

Studies have shown the additional cost of fine-tuning the PV modules' tilt angle for a roof-mounted installation to align perfectly with the Sun is not cost effective.

TABLE 16 Tilt Angle Based on Roof Slope

Roof Slope (Rise/Run)	PV Module Tilt Angle
4/12	18.4°
5/12	22.6°
6/12	26.6°
7/12	30.3°
8/12	33.7°
9/12	36.9°
10/12	39.8°
11/12	42.5°
12/12	45.0°

5.3.2 Calculating Array Size

PV system sizing starts with the customer's PV energy goals and the building's annual energy consumption.

The solar array size is determined by multiplying the annual energy consumption by the average peak sun hours (PSH) for the location and the percentage of power grid energy replacement desired by the customer.

Solar array size = annual energy consumption (kWH) × PSH × % grid replacement

For example, suppose the residential customer lives on Progress Boulevard in Alachua, Florida, with an average of 5.2 peak sun hours per day and uses 10,715kWh of energy per year. The homeowner(s) would like the solar array mounted on their roof, want a grid-tied system, and want to have enough PV modules to replace 100% of their energy usage from the power grid. This means that they would need enough PV modules to produce 10,715kWh of electricity per year.

10,715kWh per year/5.2 peak sun hours = 2,041kW/365 days = 5.59Kw

To meet 100% of the energy demand, the solar array size would need to be 5.59kW.

Number of PV Modules

The next step is to compare solar kit options. To determine the number of PV modules, use the following equation.

Number of PV modules = (solar array size/the maximum power output at STC) × a system efficiency factor

Suppose the customer finds a good deal online and wants to purchase a solar kit of six 320W PV modules. Using the specifications provided on the nameplate in *Figure 59*, would this kit generate enough electricity to meet the demand?

PV systems don't perform at 100%. Factoring in the efficiency rating of the PV modules and inverter, actual environmental temperatures, system losses, and voltage drop, PV systems are roughly 75% to 80% efficient. This equation shows how much power this solar kit would generate for the customer's location in Florida.

320W STC × 80% = 256W

6 PV modules × 256W = 1,536W × 5.2 peak sun hours = 7,987W per day or 7.99kW/day

Limits

NEC Section 690.7 states the maximum system DC voltage for a commercial building should be no greater than 1,000 volts and residential homes no greater than 600 volts. The system's DC voltage must not exceed the inverter's voltage operating limits, which limits how many PV modules can be added to the system. Refer to *NEC Table 690.7* for circuit-requirement factors.

Solar Calculators

Online solar calculators simplify system sizing calculations. For example, go to **www.sunwatts.com/solar-calculator** and plug in kWh used per year, solar hours per day, and desired percentage of solar power. It will provide general sizing information for the selected PV system.

NREL's PVWatts®

NREL's PVWatts® is a free, online solar system design tool which can model any size of array, from small residential rooftop systems to large ground-mounted power generation systems. It uses actual monthly solar irradiance data for a given location. This provides a more accurate estimate of system energy output.

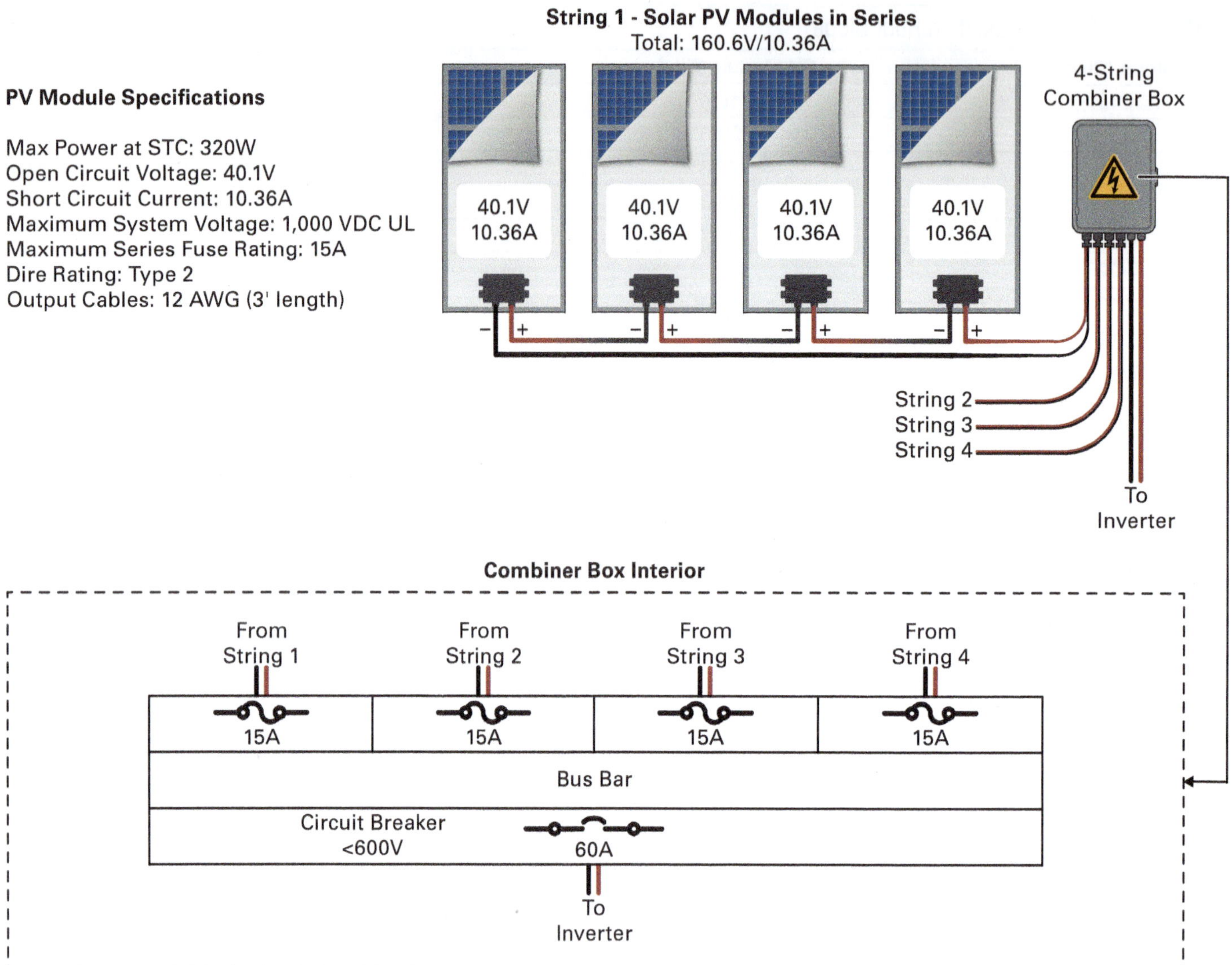

Figure 59 Calculating array size.

To determine the maximum circuit current (I_{max}) of the system, according to ***NEC Section 690.8***, the PV module's short circuit current (I_{sc}) rating for each parallel group of modules is added together and then multiplied by a factor of 1.25. I_{max} must not exceed the inverter's amperage operating limits.

For sizing overprotection, like fuses, a safety "buffer" is factored in by multiplying the maximum circuit current by an additional factor of 1.25. This factor means that the fuse size is selected to support up to 125% in the event of a power surge, for example. Refer to ***NEC Section 690.9*** for more details.

5.3.3 Inverter Selection

Inverters are designed for and selected based on the application: off-grid, grid-tied or grid-interactive, or utility-scale.

All inverters list their efficiency ratings in the product data, and higher ratings drive product selection. For example, a typical grid-tied 2,000W string inverter might have an efficiency of 95%. The inverter efficiency must be taken into consideration when sizing the inverter for the application. For grid-integrated systems, choose a hybrid inverter.

For off-grid applications, to select the appropriate inverter size, add up the wattages of all loads that are likely to operate at the same time to get the peak load. This determines the minimum continuous wattage. Next, consider the potential surge of each load to determine the minimum surge wattage. This is typically 150% to 200% of the continuous rating. For example, a 1,500W string inverter might have a 2,500W surge rating.

5.3.4 Energy Storage Solutions

Recall that grid-integrated and off-grid systems include an energy storage solution. When a power outage occurs for a grid-integrated system, it may last for a few hours. In this situation, a single battery is typically sufficient to help maintain power for that short period of time.

Off-grid systems account for 1% of the installations and look for solutions to give them three days of autonomy to cover those worst days in the year when the day is short with severe weather. In these situations, the critical loads list is used to get them through. If the storm persists, an engine-powered generator can be turned on at day four to charge up the batteries.

Main design tasks include the sizing of the battery bank, selecting a battery that meets the capacity needs, and selecting a charge controller to regulate the battery.

Battery Bank Sizing

The size of the battery bank depends on the customer's critical load list, the maximum charge and discharge rates, and the minimum temperature at which the batteries will be used. To increase energy storage capacity, batteries can be connected in series or in parallel to produce the desired output.

Wiring batteries in series increases the voltage without changing amperage. Wiring batteries in parallel increases the current (amp-hour) capacity but does not affect the voltage.

Referring to the critical loads list in *Table 15*, let's say this customer has an off-grid system and needs to size a battery bank for three days of autonomy during the winter with a critical load of 12.5kWh or 12,500Wh per day. The customer's goal is to ensure minimal heating on the potentially worst day of the year.

Table 17 and the associated image steps through the calculations in sizing the battery bank. How many lithium iron phosphate 12V 200Ah batteries are required?

CAUTION

Do not locate the inverter or batteries more than 100 feet away from the array without adjusting the wire size. Choosing the correct wire size will limit voltage drop and is crucial for system performance and safety.

Critical Load List with Versus without Heating and AC

Based on a 2022 study by the Lawrence Berkeley National Laboratory, 10kWh storage solutions with critical load lists that excluded heaters and air conditioners fully met basic backup power needs for 3-day autonomy in virtually all US counties in any month. When heating and air conditioning was added, the typical storage size increased to 30kWh.

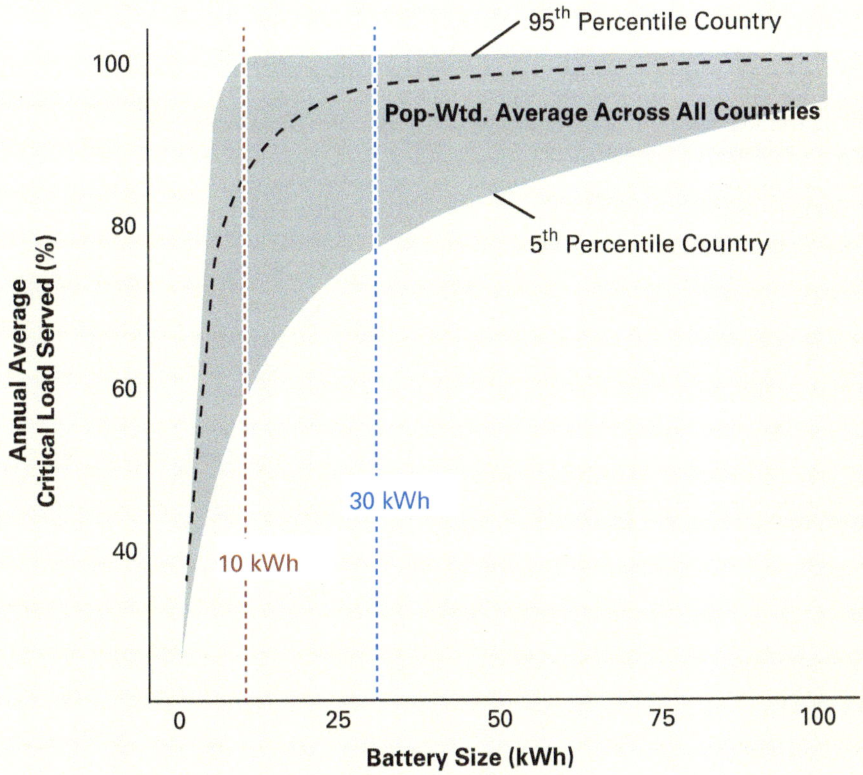

Source: UC Regents Berkeley Lab (2022), a DOE funded laboratory.

TABLE 17 Sizing the Battery Bank

Step	Math
Determine the critical load per day	12,500 watt-hours per day
Number of days autonomy	12,500 watt-hours per day × 3 days = 37,500 watt-hours for three days
Inverter efficiency rating	37,500 watt-hours per day/.92 inverter efficiency = 40,761 watt-hours per three days
Factor in the battery efficiency at stored temperature (i.e., 40°F)*	40,761 watt-hours for three days × 1.30 = 52,989 watt-hours for three days
Factor in 80% depth of discharge on the lithium-ion battery	52,989 watt-hours/0.8 depth of discharge = 66,236 watt-hours at 80% depth of discharge for 3 days Battery capacity watt-hours needed = 66,236 watt-hours

Compare Products	**12V Battery**	**24V Battery**	**48V Battery**
Use Watt's law: watt-hours/volts = amp-hours	66,236 watt-hours/12V battery bank = 5,520 amp-hours for a 12V battery bank for 3 days at 80% depth of discharge Choosing a 12V 200Ah lithium iron phosphate battery: 5,520 amp-hours/200Ah battery = (28) 12V 200Ah batteries needed	66,236 watt-hours/24V battery bank = 2,760 amp-hours for a 24V battery bank for 3 days at 80% depth of discharge Choosing a 24V 57Ah lithium battery: 2,760 amp-hours/57Ah battery = 48 batteries needed	66,236 watt-hours/48V battery bank = 1,380 amp-hours for a 48V battery bank for 3 days at 80% depth of discharge PART 1 Choosing a 48V 200Ah lithium nickel manganese cobalt battery: 1,380 amp-hours/200Ah battery = 7 batteries needed PART 2 Choosing a lithium ESS system: Manufacturer's product specifications sheet indicates 11.8kWh of usable battery capacity. ESS usable battery capacity is used to size the battery bank. 11.8kWh × 80% depth of discharge = 9.44kWh 66.236kWh/9.44kWh = 7 batteries
	Note: An MPPT feature is needed. Smart inverters have an MPPT feature available to regulate the battery bank as well. See Table 17 figure *(A) Traditional MPPT Charge Controller with Three Battery Banks.*	Exercise: Try this one on your own and review with your instructor.	See Table 17 figure *(B) Energy Storage System.*
Cost	If each battery is $400, the cost would be 28 batteries × $400 = $11,200, but fewer cycles at 50% DoD.		If the battery is $16,000, the cost would be 7 batteries × $16,000 = $112,000 with more battery cycles and higher DoD (3800 cycles @ 80% DoD).
Note	Because large battery banks are expensive, many customers scale back their critical loads list.		

*This information is obtained from a chart.

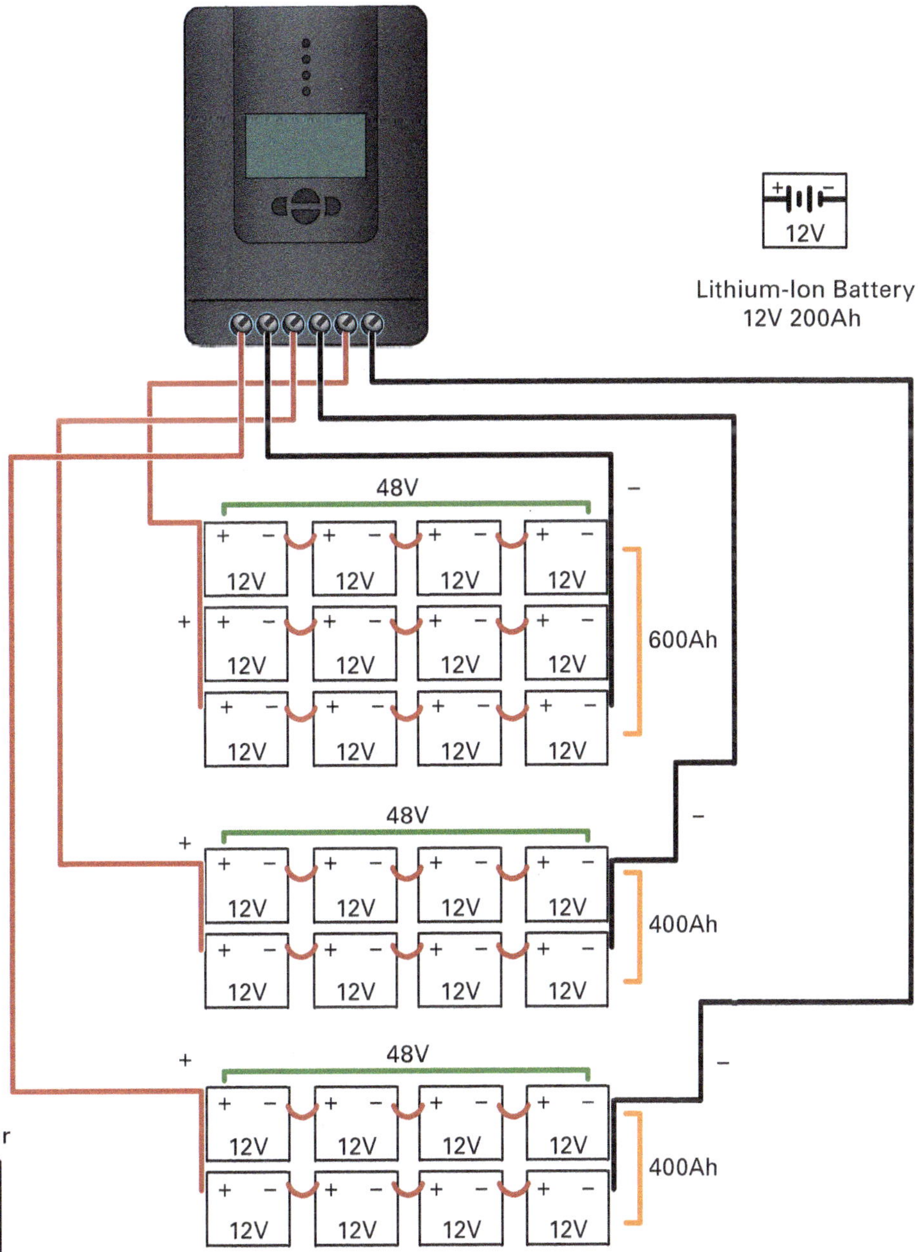

Table 17 (A) Traditional MPPT Charge Controller with Three Battery Banks

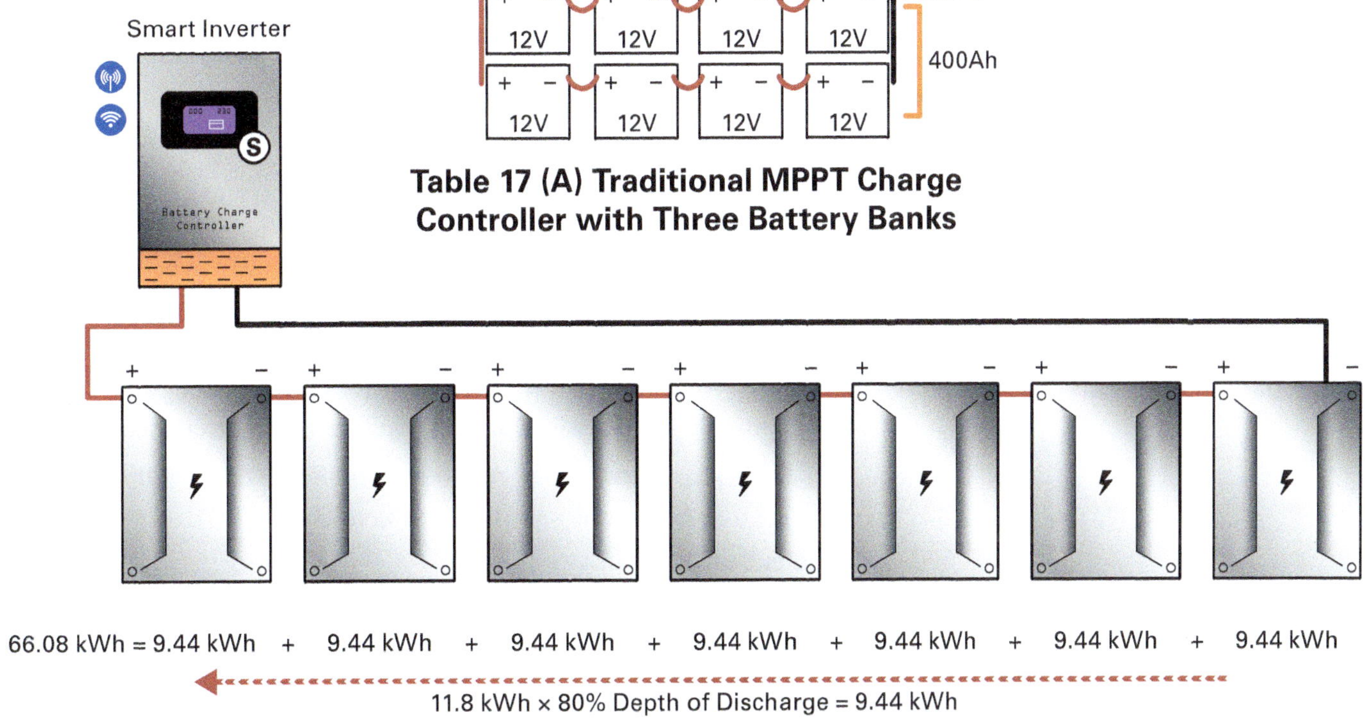

Table 17 (B) Energy Storage System

Source: (A) Based on product example from Renogy

Selecting a Charge Controller

Properly sized charge controllers help to ensure smooth power and long battery life. The selection of a charge controller depends on the application. PV systems with outputs over 200W often use MPPT charge controllers. They are more expensive but worth the investment in greater efficiency when installed to support larger systems. Smaller systems typically use PWM charge controllers.

Charge controllers must be carefully matched to the size of the battery bank (normally 12V, 24V, or 48V), the AC short-circuit current (I_{sc}) of the solar array, and the maximum system voltage. To size a charge controller, multiply the I_{sc} of the entire array by a factor of 1.25. (Note that this is a minimum value and may vary depending on the manufacturer.) The charge controller maximum input voltage must be higher than the maximum system voltage in PWM charge controllers. The design of MPPT charge controllers enables a greater nominal array voltage than the battery voltage, which permits the use of smaller wire between the battery and array.

Many charge controllers provide system output and efficiency data on a digital display.

5.3.5 Optimizing PV Conductors

PV system conductors must be adjusted for many factors, including the following:

- Temperature
- Continuous duty
- Conduit fill (more than three conductors in a raceway)
- Voltage drop

The ampacity can assist with the selection of a suitable wire gauge for the current it will carry.

Temperature Adjustment

According to *NEC Section 310.15*, conduit exposed to sunlight on a rooftop requires an adjustment for the ambient temperature. This adjustment depends on the amount of spacing between the conduit and the roof surface.

Per *NEC Section 690.7*, the maximum PV system voltage is calculated as the sum of the rated open-circuit voltages for all series-connected PV modules corrected for the lowest expected ambient temperature. The rated open-circuit voltage (V_{oc}) of a PV module is measured at 25°C (77°F). Temperature correction is required to ensure that the output voltage will not exceed the maximum input voltage of the inverter or the rating of conductors, disconnects, and overcurrent devices. The voltage correction factors are listed in *NEC Table 690.7(A)*.

NOTE

NEC Table 690.7 is based on crystalline PV modules that have been in use for many years. Newer PV modules may incorporate alternative semiconductor materials with varying temperature coefficients. Due to these variations, the manufacturer's requirements for temperature adjustment must be used where available.

The lowest available temperature for a specific area is sometimes known as the winter design temperature, and the month in which it occurs is known as the critical design month. These values can be found using local weather information or data published by the American Society of Heating, Refrigerating, and Air-Conditioning Engineers® (ASHRAE®).

To use *NEC Table 690.7(A)*, multiply the factor for the lowest expected temperature by the rated V_{oc} found on the PV module nameplate. For example, suppose a PV module has a V_{oc} of 40V and four of these PV modules will be connected in series. This system will be installed in an area where the lowest expected temperature is 0°C (32°F). According to the *NEC®* table, the correction factor for this temperature is 1.10. The system output voltage can be calculated as follows:

$$\textit{System } V_{OC} = \textit{panel } V_{OC} \times \textit{number of panels} \times \textit{correction factor}$$

$$\text{System } V_{OC} = 40\text{V} \times 4 \times 1.10 = 176\text{V}$$

Continuous Duty

PV loads are considered continuous loads. According to *NEC Section 690.8*, conductors used for continuous duty (operated more than three hours at a time) must be sized at 125% (125/100 = 1.25) of the continuous load. For example, if the actual load is 40A, the conductor must be rated as follows:

$$\textit{Continuous duty load} = \textit{actual load} \times 125\%$$

$$\text{Continuous duty load} = 40\text{A} \times 1.25 = 50\text{A}$$

Conduit Fill

More than three conductors in a raceway or cable must be adjusted using *NEC Table 310.15(C)(1)*. For example, suppose a 6 AWG USE-2 conductor is to be used with a total of four conductors in a raceway. Referring to *NEC Section 310.15*, the deduction for 4 through 6 cables in a raceway is 80% of the values found in *NEC Table 310.15(C)(1)*. A 6 AWG USE-2 conductor is rated at 55 per *NEC Table 310.16*. The adjusted value can be calculated as follows:

$$\text{Adjusted amperage} = \text{NEC Table 310.15(C)(1) amperage} \times 55\%$$

$$\text{Adjusted amperage} = 75A \times 0.80 = 41.25A$$

Voltage Drop

Excessive circuit lengths can cause a significant decrease in system output voltages. This is particularly noticeable in low-voltage PV systems. In PV systems, voltage drop should be limited to 3% or less.

5.3.6 Forces Exerted on the PV Modules and Support System

Solar arrays are subjected to a variety of forces that can stress the PV modules and structural components. These forces include expansion and contraction, drag, and wind. Consult a licensed structural engineer for performing a load analysis.

Expansion and Contraction

As a PV module heats and cools, the metal frame and array mounting structure expand and contract. This change in length is multiplied when PV modules are racked side-by-side in a large array. For this reason, any installation more than 50 feet in length should include an expansion joint for both the array and the mounting structure or as designed by the professional engineer (PE) of record. The PV module expansion joint should not straddle the mounting PV module expansion joint.

Snow and Ice

If an array is installed in an area with heavy snow and ice loading, the PV modules will be subjected to a drag force that pulls the PV modules at a downward angle. To resist the drag force, the PV module must have support on the lower edge of the PV module frame. Additional mounting clips may be required for systems with excessive drag.

Wind

Wind loads can stress or damage PV modules and supports via drag or uplift forces. In windy conditions, a solar array can exert several thousand pounds of uplift force against the structural supports. Wind loads are particularly critical with angled roof supports because the PV modules can act as a sail, pulling the array away from the structure. Wind can loosen fasteners over time.

Careful evaluation of wind loads is essential to ensure a safe installation and is also required to secure certain types of rebates. Wind loads are labeled on support systems and their associated fasteners. Some pre-engineered (PE) systems may be supplied with a PE stamp to indicate that they have been evaluated for a specific wind load. All installed systems require PE certification. In addition, systems installed in earthquake-prone areas must be rated to resist seismic loads.

Permits: SolarAPP+

To process solar PV permits efficiently, NREL has developed an online tool with inspection forms for data entry. This tool, SolarAPP+, is intended to speed up the review and approval process to install solar PV.

5.3.7 Proposals

The system design, sometimes with iterations for product and performance comparisons, and electrical plans are compiled into a proposal to present for customer approval, financing, and permits.

5.4.0 Installation

Once site assessment and system design are completed, a solar PV system will be ready for installation. This section provides an overview of system installation, including roof- and ground-mounted installations as well as electrical installation basics. Solar PV installers must also understand the foundations of system maintenance and troubleshooting that come into play after a system is installed.

An important part of installation is organizing the jobsite and managing team communications. Site management verifies the job permits are posted and establishes a jobsite analysis to communicate hazards and remedies at the site. Fall protection, scaffolding, and barriers (*Figure 60*) are set up by workers who have certified competency in those areas of expertise. Special equipment, such as aerial lifts or cranes, is scheduled to be at the jobsite with certified operators on the day of installation.

Barrier System for Edge of Building

Figure 60 Fall protection example.

Think About It

Roof Age and Damage

When you look at the condition of the roof in *Figure 60*, would you start installing a PV array on that surface or would you have a conversation with the building owner first about roof replacement or maybe repairs?

Access to the site and orderly placement of equipment and materials make the job run smoother. For ground-mounted installations, underground utilities must be marked.

For housekeeping, in general, pick up after yourself and be mindful of others. Keep walkways clear of tools and unused equipment. Keep tools leashed and materials and equipment contained for the safety of others, especially on sloped roofs.

5.4.1 Team Communications

Safety is a top concern at the jobsite. Required PPE is discussed, and the group reviews the job safety analysis (JSA). The JSA process identifies jobsite hazards, recommends remedies, and communicates the concerns to the team.

A jobsite may have a trailer where workers can check out specialized tools. The location of this trailer, security measures, and the basic plan of where parts are delivered and stored are all explained at the start of the job in a team meeting.

A job packet provided to workers contains some of the site assessment report details, PV module layout and electrical schematics from the solar system design, and manufacturer product details. Use the schematics to look up wiring details and placement of the PV components. Read through all this information to understand the solar photovoltaic (PV) installation plan.

Additional training is provided by the employer.

5.4.2 Prepare the Site

Depending on the size of the job, some organizational tasks may be completed before the installation.

The solar system design software provides a Bill of Materials, which is a parts list used for the following purposes:

- Purchase and inventory of parts
- Estimating the size of the job to establish support vehicle requirements and designated areas for specific tasks

In some cases, the Bill of Materials is printed on the drawings. *Table 18* shows common elements of a Bill of Materials.

Material Delivery and Handling

Designate work areas, or *laydown areas*, and clear them of as many obstacles as possible. Organize the flow of vehicle traffic and the unloading of equipment and parts.

Secure PV components in nearby storage and take care not to over-stack palettes of PV modules. Broken PV modules are expensive to replace and may create delays in receiving new parts. For safety, PV modules are among the last items installed, and before installation, they are electrically tested. Using a Bill of Materials, inventory all materials, label them to make the task of identification and assembly easier, and store them in their designated places.

Protect PV components within their shipping materials until they are needed. Inverters ship with shock indicators to identify when the equipment may have jarred enough to damage the unit. Some of the structural components are long and bulky, so centrally locate the pickup point to balance the load.

Packaging materials need to be contained and properly disposed of after they are removed.

NOTE

Kitting PV module attachment parts is a time-saver task that increases efficiency of workers. Assembling clamps and hardware parts in preparation for installation reduces the trips to find things and minimizes assembly times in the field.

General Installation Steps

Up until now, most of the content has begun with the PV module. However, in installation, this is the last item to be addressed because the minute the surface is exposed to sunlight, it begins to generate electricity and poses a safety risk.

TABLE 18 Materials List for a Solar PV Installation

Items	Model	Quantity	Remarks
PV modules			
Optimizers			
Combiner boxes			
Inverter(s)			Type: _____
DC disconnects			
AC disconnects			
Charge controller(s)			
Batteries			Type: _____
Metal raceway			10' lengths of 3/4" EMT
90-degree elbows			3/4" EMT
45-degree elbows			3/4" EMT
Connectors			3/4" EMT (insulated and rain tight)
Couplings			3/4" EMT (rain tight)
Wire from array to charge controller			Type: _____ Length: _____ Installation rating: _____
Wire from charge controller to batteries			Type: _____ Length: _____ Installation rating: _____
Wire from charge controller to inverters			Type: _____ Length: _____ Installation rating: _____
Wire from inverters to AC breaker			Type: _____ Length: _____ Installation rating: _____
AC breaker			

For the safety of you and your crew, and assuming that the mounting surface has been prepared, the general order in which the PV components are assembled is as follows:

1. Install mounting hardware and raceways
2. Install PV components
   A. Combiner
   B. Optional energy storage
      i. Batteries
      ii. Charge controller
   C. Inverter
   D. Safety components
   E. PV modules
3. Install wiring

5.4.3 Ground-Mounted Considerations

Soil studies are conducted which guide design decisions about what type of foundation is used for the environmental conditions at the jobsite. Regardless of the foundation, the assembly steps for a fixed-tilt system are the same.

Civil engineering design software with Global Positioning System (GPS) technology is used throughout the construction process as models guide vehicles in excavating the site according to the site plans.

Local utility companies identify and mark utility locations.

Mark Utility Locations

Underground utilities must be marked (*Figure 61*). If the installation site is on the roof of a building or home, the same lines need to be clearly identified around the structure. Aerial lifts or trucks hauling in the materials can damage these lines if they run over the ground above the lines. Overhead lines are also a concern.

CAUTION

If the installation site is out in a field, there may be digging tools in the area. Exercise caution when walking around a field installation site. The ground may be uneven, which may cause either a tripping accident or a loss of balance.

WARNING!

All states require that the local utilities be contacted before digging. The "811 Call Before You Dig" program is available throughout the United States. Learn more at https://call811.com.

NOTE

This introductory module is designed to give you a basic overview of solar installation. For more details on solar system design, consult NCCER Module 57108, *Ground-Mounted Solar PV Installation*.

Figure 61 Markers indicate the location of pipeline utilities.
Source: Grb/Getty Images

Earthmoving

Excavation of an area (e.g., grading, leveling, trenching) may be required to prepare the earth for foundational work. A utility locating company establishes the location of underground utilities. If holes have been dug for foundations, make sure that they are clearly marked with some kind of barrier (*Figure 62*).

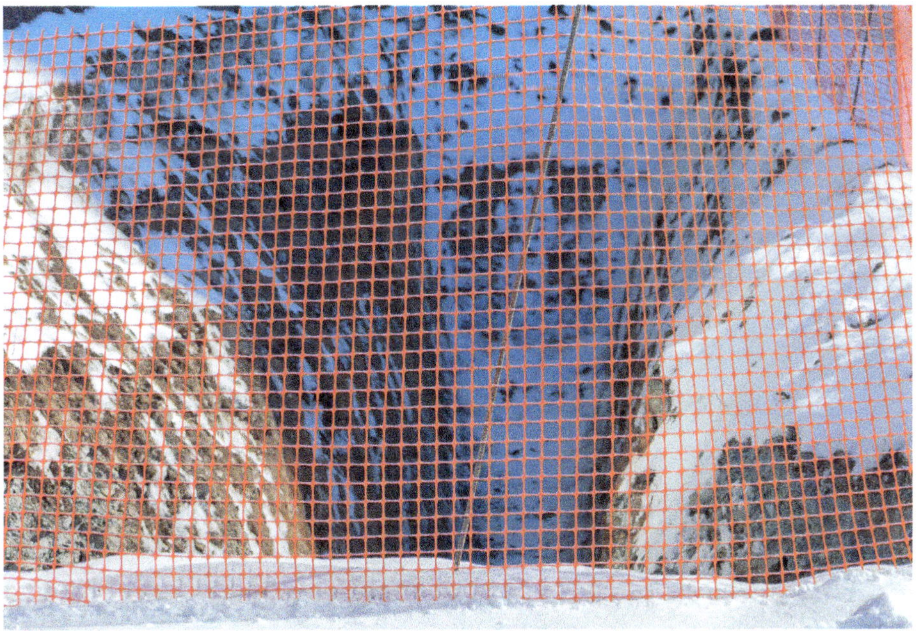

Figure 62 Example warning barrier for site trenching and excavation.
Source: vladimir subbotin/Shutterstock

Racking System Installation

There are several methods used to install ground-mounted systems. Whether the footers are constructed out of concrete or metal posts driven into the earth, a racking system is erected to secure the PV modules.

Follow the system design's planned PV module tilt angle and azimuth angle positioning with the Sun.

The following steps outline a general installation procedure for a ground-mounted array:

NOTE

Sometimes the soil isn't suited for drilling, and a ballast-mount installation has been planned.

Step 1 The site is cleared of any vegetation and graded.

Step 2 Concrete footers or metal posts are installed to support the array's racking system (*Figure 63*).

Step 3 The racking system is erected onto the footers according to the manufacturer's instructions (*Figure 63*).

Step 4 PV modules are tested, installed, and wired.

All fasteners must be tightened to the manufacturer's specifications. This is known as *torque*. Be sure not to under- or over-torque your fasteners (*Figure 64*).

Wiring and Raceways

Most utility-scale PV installers "pull" wires and cable to termination locations in preparation for a certified electrician to make the connections. A completed

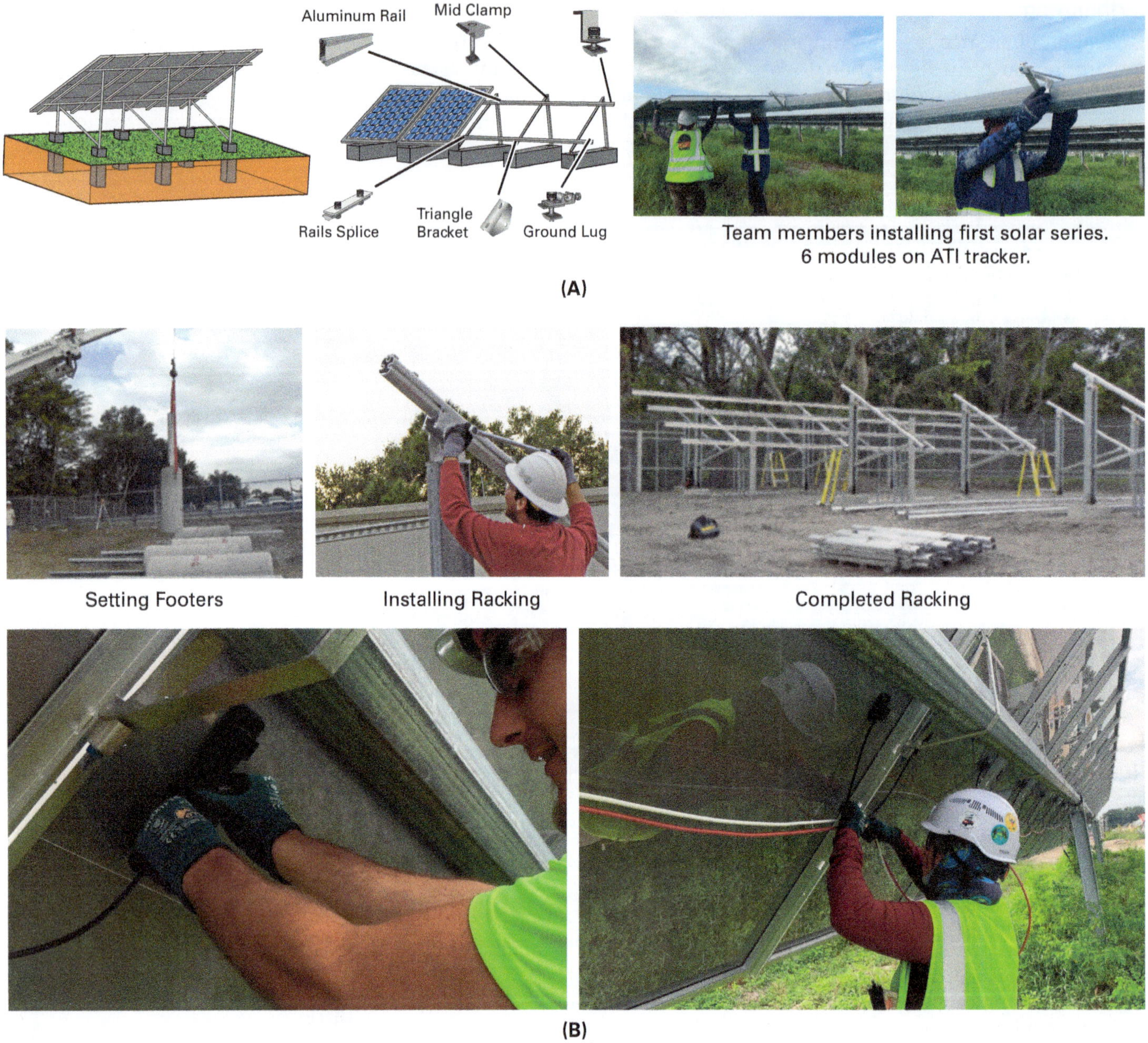

Figure 63 Support system and PV module installation examples.
Source: Mortenson

Figure 64 Torquing fasteners.
Source: JPSchrage/Getty Images (bottom left)

ground-mounted installation is shown in *Figure 65*. The image on the right gives an idea of the amount of wiring that may be required for a job.

Ground-Mount Installation with Post Footing

Organizing Wires

Figure 65 Utility-scale ground-mounted installation.
Source: milehightraveler/Getty Images (left); Tsvetan Ivanov/Getty Images (right)

5.4.4 Roof-Mounted Installations

Recall that there are low-slope and steep-slope roofs and review the PPE for each situation. For roof-mounted installations, the site assessor reviewed the integrity of the roof. Any structural problems should have been addressed prior to the start of the solar PV installation.

In general, disturb the roofing surface as little as possible, and follow the system's design specifications. Follow the manufacturer's instructions for module spacing and fastener torque specifications, and electrically test each PV module before installation.

Think About It

Fastener Checks

Some installers use a permanent marker to draw a line over the center of the fastener when it is initially installed and after it has been checked for proper torque. That way, they can tell at a glance when fasteners have become loose. Can you find the loosened fastener in this picture?

Source: Mike Powers

Roof PV modules can be laid out in either a portrait (vertical) or landscape (horizontal) orientation. Be sure to consult your design specifications to ensure you install panels in the correct orientation.

Any roof penetrations to anchor support systems require flashing and weatherproofing.

Low-Slope Installation

Follow the system design's planned PV module tilt angle and azimuth angle positioning with the Sun.

A low-slope solar PV installation may secure PV modules to the building using ballast mounts, direct attachment, or a combination of both as follows:

- *Ballast mounts* — The framing structure is assembled to manufacturer's specification and the concrete blocks are set into position. Over time, the weights shift and may tear the roof's membrane. Mats may be installed under ballast trays to minimize damage to the roof.
- *Direct attachment* — This installation type directly anchors a racking system to the roof to support the PV modules.
- *Hybrid* — This installation type combines ballast mounting and direct attachment to reduce weight and improve wind load factors. Some anchor points may be installed.

Steep-Slope Installation

For steep-slope installation, an approved fire-rated roofing system must be installed on the roof deck before installation of a PV array. Selection of the roofing system depends on local code requirements, building design, and user preference. In many cases, the fire-rated system consists of asphalt shingles or concrete roof tiles. It must cover the entire roof surface, including areas where PV modules will be installed.

PV modules are aligned with the roof's slope. Additional tilt angle optimization is not cost effective. To minimize shading losses at low sun angles, follow the manufacturer's requirements for PV module spacing. Aesthetics is part of the job, so center the array on the roof.

CAUTION

All roof penetrations, including fastener heads, must be weather sealed to avoid leaks.

To ensure system integrity, all mounts must be attached to structural members such as rafters and trusses.

For a steep-slope roof, PV modules are secured to the roof deck in one of three ways: direct mount, stand-off mount, and building-integrated PV.

For metal and wood surfaces, mounting hardware is anchored directly to the roofing rafters or trusses. In some situations, additional spans of 2″ × 6″ wood sections can be installed between the rafters or trusses. These extra spans provide additional anchorage for the PV module mounting attachments. Follow the manufacturer's requirements for anchor points and spacing.

5.4.5 General Installation Steps for a Rack Mount System

Install a rack mount system as follows:

Step 1 Find the locations in the structure where the mounts can be attached.

Step 2 Open the roof.

Step 3 Attach the mounts.

Step 4 Flash and close the roof.

Mounting hardware varies widely by manufacturer. During the design phase, the designer determines the number of anchor points and mounting hardware items needed for the given location with weather conditions in mind.

CAUTION

Special considerations are taken for installation on tiled roofs.

Plan the layout of the mounting hardware in relation to the mounting surface.

Following the design, the roof layout is measured and chalked on the roofing surface. Chalk lines help visualize the location of the array, the rows, and the attachment points. The first anchor point is established, and the rest are measured and marked with chalk from there. For wood construction, the first anchor point is established from the interior by drilling a hole beside the first rafter or truss. Then, from the exterior, the on-center distance from rafter to rafter is measured and chalked, establishing the placement of the anchor points to be drilled.

To install the mounting hardware, anchor points are drilled into the mounting surfaces, which break, or penetrate, the roof's weatherproofing. Without resealing, the roof may leak. The number of required roof penetrations depends on the following factors:

- The roofing structure and manufacturer recommendations
- Solar PV system design
- Local building codes
- Weather conditions for the location, which may increase the number of anchors

For concrete decks, anchor bolts may be driven into the concrete (*Figure 66*).

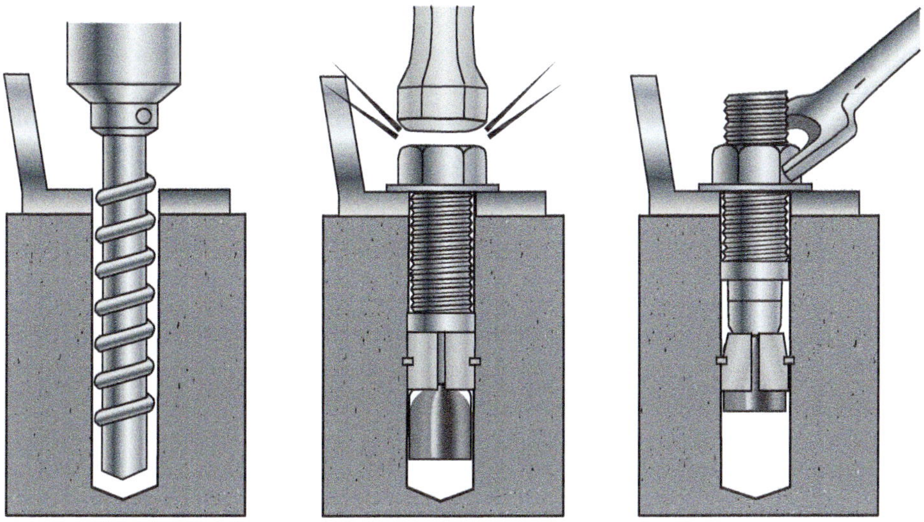

Figure 66 Concrete anchor bolts.

Regardless of the surface type, all mounting holes needed to install the PV system must be pre-drilled to the proper pilot hole size for the hardware being used. Follow manufacturer's instructions and details for proper flashing and weatherproofing around all mounts and close the roof.

Attach the rack system to the anchors. Run the grounding wire. Wire the PV modules and clamp them to the rails.

5.4.6 Electrical System Installation

After the array has been installed, the final electrical connections are made. The PV modules can be joined using traditional conductors and MC4 connectors. The factory wiring is identified for the positive (+) and negative (–) PV module output polarities.

WARNING!

MC4 connectors vary by manufacturer. To reduce the chance of an electrical arc, use MC4 connectors from the same manufacturer throughout the PV system.

Grounding

Size and install the equipment grounding conductor in accordance with local requirements and the *NEC*® guidelines. Attach the equipment grounding conductor to the PV module frame using the hole and hardware provided.

The roof ground should be a minimum 6 AWG solid copper to provide extra strength. Always use approved cleats when making ground connections. They are designed to pierce the finish of the panel to ensure electrical continuity.

WARNING!

Always observe proper precautions when connecting or disconnecting PV modules exposed to light, because hazardous voltage may be present. Note that quick-connect termination boxes contain no user-serviceable components. Do not open the boxes or alter the wiring in this type of system.

Wiring

Connections are also made to the inverter, batteries and charge controller (if used), PV module, and disconnects. Use USE-2 (sunlight resistant, 90°C), for all wiring exposed to weather. The minimum wire size is listed on the PV module nameplate. For example, a PV module might list the minimum wire size as 12 AWG and the maximum wire size as 8 AWG. (Remember, wire sizes increase as the AWG size gets smaller, then they jump to 1/0, 2/0, and so on.) Most PV wire is 10 AWG.

Grid-tied systems typically feed into the last two spaces at the bottom in the main service panel, the furthest spaces from the main feed. When adding power to a PV module, do not exceed 120% of the bus rating. See *NEC Section 705.12(B)(3)*. Additional capacity may be added, when necessary, by reducing the size of the main breaker or tapping ahead of the bus feed and adding a fused disconnect. Label the PV components according to the *NEC*® guidelines. *Figure 67* shows a completed electrical installation.

Figure 67 Components in proximity to one another.

5.4.7 System Commissioning

The final step is to commission the solar PV system, which authorizes the use of the solar PV system. All parties involved must agree and they may require some of the following procedures:

- Roof inspection
- Electrical inspection
- Fire inspection
- Structural or building inspection
- City/county permit approval
- Utility company permit approval and interconnection agreement

Using a checklist, the crew systematically inspects the PV system. For example, fasteners are randomly spot checked within the entire system to ensure that 2% to 10% of the fasteners are properly torqued to a manufacturer's specifications. If the check fails, then check all fasteners in the affected rows.

The utility company inspects the system, testing it to ensure that safety standards and energy output performance are met. Check for output voltage, current, and polarity. Program inverters and charge controllers. Review the operation of the system with the building owner and provide all operating manuals for the installed equipment.

Assessing System Output Power

Assessing the expected output of a grid-tied PV array requires the application of various PV module correction factors. Typical optimization factors include the following:

- Solar irradiance
- Array temperature and ambient temperature at the time of testing
- PV module mismatch and production tolerances
- Dust
- Wiring losses
- Inverter conversion losses

Taking into account all of these system losses, the actual array output should be about 70% to 80% of the STC value. This means that the measured output of a 1,000W PV system should be somewhere between 700W and 800W on a sunny day.

Assessing the expected output of a grid-interactive system is more complicated due to the addition of the batteries.

5.5.0 Maintenance and Troubleshooting

While PV systems generally have low maintenance requirements, performing recommended tasks, known as operations and maintenance (O&M), at the recommended intervals maintains system performance. Keep maintenance and service records.

NOTE

This module is designed to give you a basic overview of solar installation. For more details on maintenance, consult NCCER Module 57105, *Maintenance and Troubleshooting*.

WARNING!

Only qualified individuals may perform PV system maintenance tasks. Wear all the required personal protective equipment when installing PV systems. Apply sunscreen and stay hydrated when working in hot temperatures. Always use fall protection on work surfaces six feet or more above the ground. Do not step on, walk on, or rest objects on PV modules. Be aware of all electrical hazards.

Solar monitoring software systems or drones with thermal cameras may be used to monitor the energy output of the solar PV system. When the energy output performance declines, it is most likely time to clean the PV modules, although dirt may not be the only cause. To clean, use the manufacturer's recommended cleaning solution.

Make seasonal adjustments to the array's tilt angle, if possible.

CAUTION

PV modules may be damaged by high-pressure water, hard bristled brushes, or any types of solvents, abrasives, and harsh detergents.

5.5.1 PV Module Performance

For electrical equipment inspections, follow the equipment manufacturer's recommendations. Test the performance of the PV system before and after cleaning. Assess the following:

- *Energy Production* — Measure the amount of DC energy that the PV arrays are generating in kWh. Compare this value with the PV module's nameplate DC peak power rating (STC) and site-specific solar irradiance measurements.
- *AC Energy Output* — Measure the inverter's AC energy output and compare the value with the output expected for the system.

Cleaning Robots

Flat mounted solar farms by Erthos® have optional robotic cleaning service plans available.

Source: Erthos

Energy Production

A pyranometer (*Figure 68*) is mounted in plane with the solar PV array and measures real-time solar irradiance. It creates a realistic expectation of the amount of energy that is possible to produce at that location with readings in watts/m^2. Over time this data is compared against the actual solar PV array energy output to assess the solar PV system performance.

Handheld solar irradiance meters (*Figure 68*) are less precise but offer the advantages of portability and rapid field measurements.

If a pyranometer is not available, NREL's online tool PVWatts® can be used to determine solar irradiance for a location. Using regional solar irradiance data for a given location, PVWatts® estimates average power output and energy production for a year.

Once peak sun hours have been determined from the actual solar irradiance, then the system's performance can be calculated. The average daily AC energy production (AC kWh) divided by the product of the PV array DC peak power rating at STC (DC kWp) and peak sun hours (PSH) is a key indicator of system performance.

The system never performs at 100% capacity due to temperature and system losses. The average is between 70% to 85%. So, if you chose 85% or 0.85, the equation would look like this:

$$\text{AC kWh}/(\text{DC kWp} \times \text{PSH}) = 0.85$$

(A) Pyranometer **(B) Solar Irradiance Meter**

Figure 68 Measuring solar irradiance on-site.
Source: (A) behindlens/Shutterstock; (B) Fluke

Voltage Output Levels

One performance test is to check the inverter's AC power output.

Check the PV system's grounding equipment for continuity to the earth in accordance with the *NEC*® guidelines. On the utility scale, multiple feeder connections are managed in a single unit called a solar switchgear compartment. One maintenance task is to check the switchgear compartment. The electrician in *Figure 69* is using a Fluke 393 FC CAT III 1500V true-rms solar clamp meter.

Figure 69 Utility scale: Checking the switchgear compartment.
Source: Fluke

5.5.2 Troubleshooting

A performance drop in the PV system, either at the array or the inverter, indicates a problem. Many PV components have software monitoring and reporting features. *Table 19* lists common problems and remedies.

NOTE

Over time, PV modules degrade and naturally have a reduction in DC energy output of about 0.5% per year for high-quality PV modules.

5.5.3 Visual Inspections

Perform a visual inspection of the PV modules. Look for problems such as dirt, bird droppings, and cracks in the PV module glass. Audit the fasteners to tighten them back to the manufacturer's torque specifications. If you find problems, perform tests to identify and isolate system malfunctions.

WARNING!

Broken PV module glass is an electrical safety hazard (electric shock and fire). These PV modules cannot be repaired and must be replaced. If you find a broken PV module, turn the system off until the PV module has been replaced.

Thermal imaging is used to troubleshoot solar PV system modules and wiring (*Figure 70*) and battery banks (*Figure 53*). The images can be taken by drones with thermal cameras or by handhelds. Smartphone camera adapters are available as well. Delamination may occur around the edges of a PV module, which creates an entryway for water, among other things.

TABLE 19 Common Problems in PV Systems

Problem	Observation	Cause	Remedy
Inverter loss	Drop in the inverter AC output that is not due to shading	Dust/dirt build up on PV modules	Clean the PV modules per manufacturer's instructions
Faulty PV module wiring	Thermal imaging shows cool colors for sections of wire	Damaged wiring; rodent chewing wires	Look for a mistake in the wiring or disconnected wiring
Heat fade	Confirmed by spraying water on the PV modules to cool them down while monitoring current fluctuations	Poor connections, undersized wiring, or extreme environmental temperatures	Increase, or *upsize*, the wire gauge If performance does not improve, then the PV module may have been damaged by excessive heat
Burnt terminals	Discoloration of the terminals	Repeated temperature cycling, oxidation, corrosion, or too many PV modules in the same circuit	Replace all metal parts that have been severely oxidized; rewire some PV modules to another circuit
Bypass diode failure	PV module voltage drop	Voltage surges (lightning)	Replace the entire PV module per the manufacturer's instructions
Delamination	Water entry around the edges of the PV module; gaps between the glass	Weathered surfaces, high temperatures, manufacturing defect	Replace the entire PV module per the manufacturer's instructions
PV thermal event	Electrical arc flash	Mismatched MC4 connectors	Do not mix MC4 connectors from different manufacturers
Oxidation	PV module turns yellow or brown	Quality of materials chosen by the manufacturer; environmental conditions	Replace the entire PV module per the manufacturer's instructions

*This is not an exhaustive list.

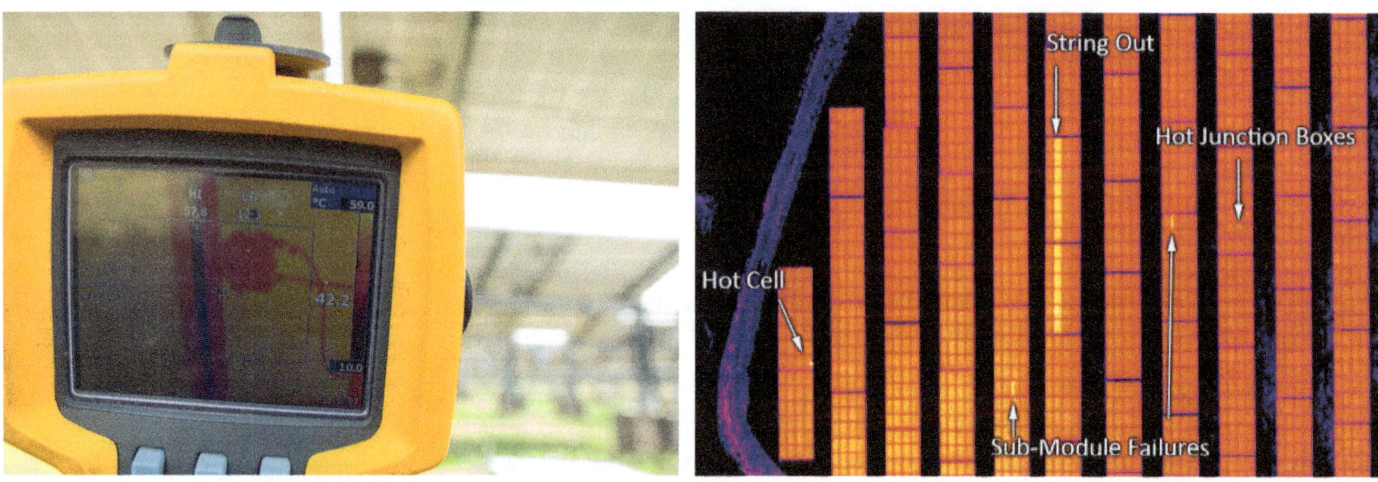

(A) Wire Checks **(B) Aerial Solar Array Checks**

Figure 70 Thermal imaging assists with maintenance.
Source: (A) undefined undefined/Getty Images; (B) NREL

5.0.0 Section Review Questions

1. Customer utility bills establish the building's energy consumption to help _____.
   a. identify energy efficient products
   b. provide a tax incentive
   c. reduce carbon dioxide emissions
   d. size the solar PV system

2. San Antonio, TX, has 5.25 peak sun hours per day. If you were estimating the energy output of a 6kW residential system, how many kilowatt-hours would be generated per day?
   a. 22.5kW/day
   b. 25.5kW/day
   c. 30kW/day
   d. 31.5kW/day

3. All installed PV systems require a professional engineer (PE) stamp indicating the rated _____.
   a. expansion differential
   b. drag
   c. wind load
   d. snow load

4. Which of the following steps is unique to ground-mounted system installations?
   a. Wire, ground, and bond the PV modules.
   b. Test and install PV modules.
   c. Clear the site of any vegetation and grade.
   d. Erect the racking system.

5. Which of the following steps is unique to roof-mounted system installations?
   a. Chalk array position, drill pilot hole, measure and mark attachment points.
   b. Test and install PV modules.
   c. Wire, ground, and bond the PV modules.
   d. Erect the racking system.

6. Which of the following issues is most likely to be observed if a PV system operates inefficiently during periods of high heat?
   a. Heat fade
   b. Burnt terminals
   c. Oxidation
   d. Inverter losses

6.0.0 Solar PV and the Power Grid

Objective

Understand fundamental changes happening to the power grid in response to renewable energy additions into the system.

a. Explain power generation and distribution trends.

Performance Tasks

This is a knowledge-based module. There are no Performance Tasks.

Utility-scale solar generating systems will support the new infrastructure of public and private EV charging stations that is being developed to support the growing US energy demands. EV charging stations are already placing demands on the current system. A **brownout** occurs when demand on the power grid exceeds its capacity. This temporary decrease in grid output voltage can result in dimmed lighting and harm electronics. As more commercial and residential solar PV systems, also known as **distributed energy resources (DER)**, connect and provide electricity to the power grid, the occurrence of brownouts may be reduced.

Safety regulations continue to develop to support the health of the power grid, as the voltage within the regional systems climbs. Guidelines such as the introduction of anti-islanding smart inverter features prevent backfeeding power into the grid.

In the United States, the power grid is divided into three main regions called the Eastern, Western, and Texas Interconnects, which can then be divided into

Utility-scale solar generating systems: Large solar farms designed to produce power in quantities large enough to operate a small city.

Brownout: A temporary decrease in grid output voltage typically caused by peak load demands.

Distributed energy resources (DER): Independent, renewable energy source generators, such as residential solar PV systems.

smaller regional groups. Upgrades across the entire system happen regionally, making some sections more technologically advanced than others.

Traditionally, the power grid has three main parts: generation, transmission, and distribution of energy. However, Smart Grids are power grids that have a plant control software system that communicates to smart devices within the network. This leads to a unified and simplified system for control, maintenance, and management of the power grid, including generation, transmission, distribution, storage, and trade of energy.

6.1.0 Management of the Power Grid

A lot of equipment and many computers and people all work together to keep the power grid operating. A utility company's central computer system securely monitors the distribution of energy between commercial and residential customers. Cybersecurity is a fundamental piece of operations, but beyond the scope of this module.

Software solutions use wireless or cellular networks to connect the customer's Home Area Network (HAN) to the Neighborhood Area Network (NAN). This software enables smart devices to monitor and report system performance to central utility controllers. For example, a smart inverter reads and reports system voltage and amperage, which the utility company uses to regulate the safe flow of the energy into the power grid. A building's smart meter reads and reports the power consumed to the utility company and may be shut down remotely during emergencies.

6.1.1 Power Generation

Power generation comes from several energy sources, including fossil fuels, hydro, solar PV, solar thermal, and wind, which are also connected to the power grid. *Figure 71* (A) shows the plan for the United States to retire some energy solutions, such as coal (gray) and nuclear (red), *while* (B) shows how more renewable energy sources, especially solar PV (yellow) will be added in the future.

Traditional power plants and distributed energy resources, including residential and commercial solar PV installations that participate in net-metering programs, generate electricity from solar energy.

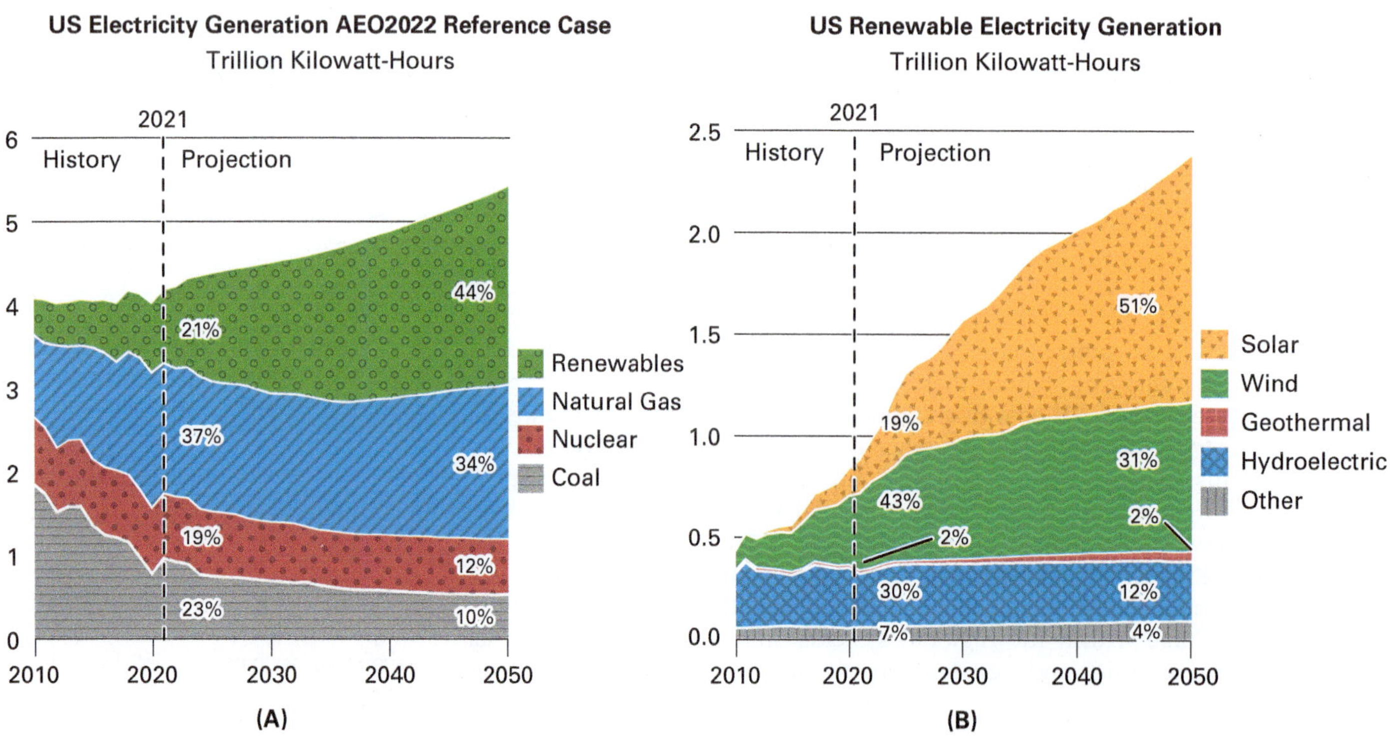

Figure 71 Projected US electricity generation by source.
Source: NREL

6.1.2 Power Transmission

Transmission refers to the parts of the network that carry high-voltage AC electrical power over long distances from a power plant to a distribution substation, which distributes the electrical power to the consumer (*Figure 72*). Substation operators monitor consumer energy demands and route supplies as needed.

Transportation across Long Distances

For the utility scale of operations, you can imagine that the wiring and voltage requirements are vastly different from those in your home.

Electricity is transported long distances and the voltage levels vary along the line. Step-up transformers increase the voltage and step-down transformers reduce the voltage. *Figure 72* shows the voltage levels along the system from the generating system down to a residential building. Recall that residential buildings have 120V for lighting and small appliances and 240V to run major appliances such as clothes dryers.

Three-Phase and Single-Phase Wiring

Three-phase wiring is designed to carry constant, steady, high voltage AC power across long distances. It is composed of three or four aluminum wires, with the wires aligning the current in a phase 120° from each other. One of these wires is a neutral. Large commercial buildings and factories also use three-phase wiring.

Single-phase wiring is used in residential and small business wiring configurations and supplies AC power up to 1,000 watts. The current flows along one power wire and one neutral wire.

Power Plants

Power plants output varying voltages. A typical plant output is between 13,000 volts and 14,000 volts. While this seems high, it is not enough to travel across miles of transmission lines without significant losses. To overcome voltage losses, the voltage is stepped up to an even higher voltage before it leaves the power plant.

NOTE

Not all utility cables are above ground. Underground distribution systems have advantages: for example, protection from weather and space efficiency in urban areas. They also have disadvantages: they are costly, subject to flooding and corrosion, and harder to repair.

6.1.3 Power Distribution

Traditionally, all electricity was generated by the utility company with one-way distribution to buildings. Now, the power grid, section by section, is being updated to enable the flow of electricity to and from the power grid to support the addition of solar PV systems. These updates include electrical safety and computer security measures.

As described in *Figure 73*, the grid's distribution components include the interconnections, power poles, step-down transformers, power lines (wires), sensors, step-up transformers, and substations.

Points of interconnection for the building with the utility company's power grid include the building's electric meter, smart meter, or smart inverter. Some buildings may have more than one meter. For example, a hospital may have a dedicated meter for refrigeration.

NOTE

This module is designed to give you a basic overview of solar photovoltaics. For more details on upgrading at the utility scale, consult NCCER Module 57105, *Maintenance and Troubleshooting*.

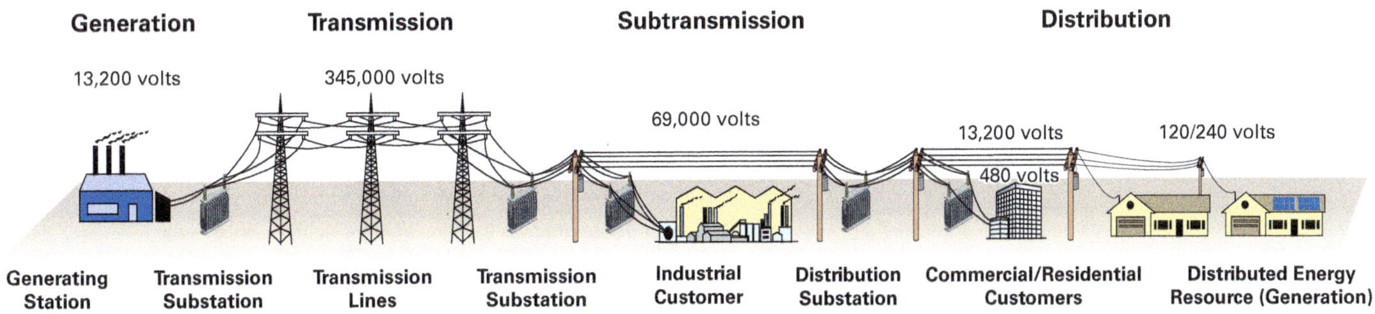

Figure 72 Utility-scale system.
Source: NREL

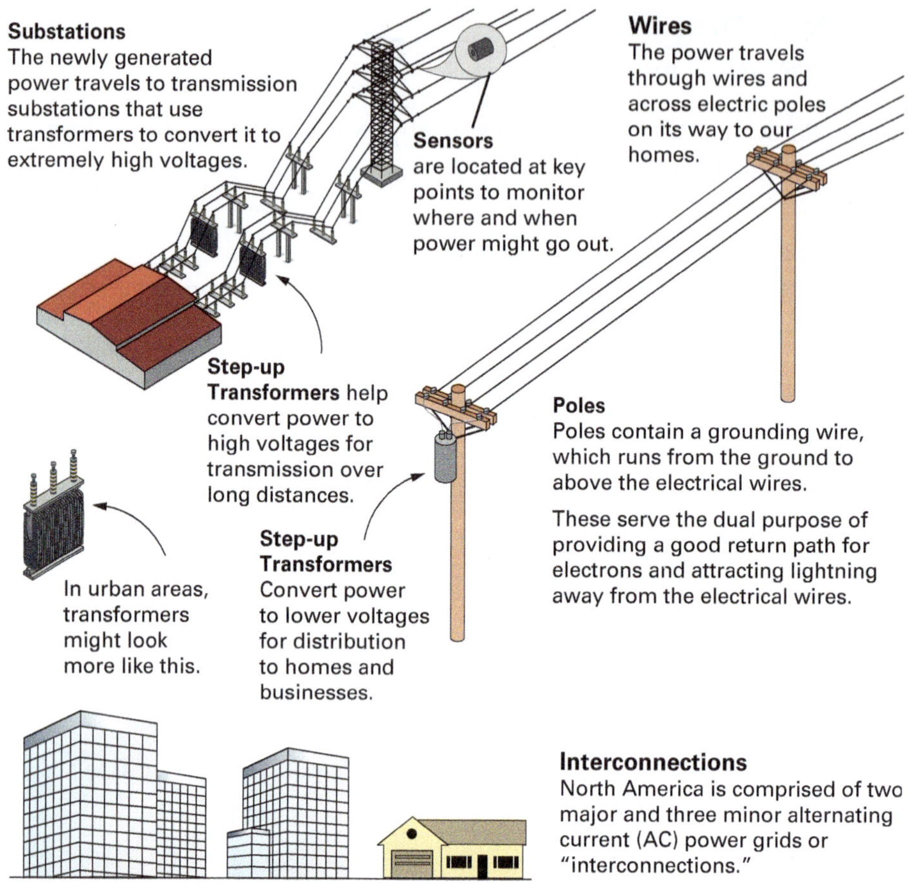

Figure 73 The power grid's subtransmission and distribution subsystem.
Source: Based on information form the U. S. Department of Energy

6.1.4 Storage

As solar PV energy is added to the grid, new energy storage solutions will need to be developed to increase energy storage capacity (*Figure 74*) for distribution of that energy at any time of day.

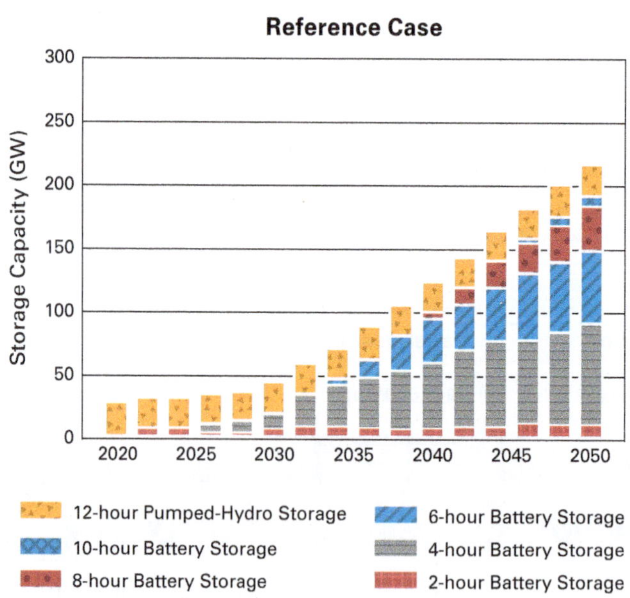

Figure 74 US energy storage capacity.
Source: NREL https://www.nrel.gov/docs/fy21osti/77449.pdf

6.0.0 Section Review Questions

1. A wireless internet or cellular connection is required for which of the following devices?
   a. Charge controller
   b. Service panel
   c. Smart inverter
   d. Transformer

2. Which energy solutions does the United States plan to retire?
   a. Coal and hydro
   b. Coal and nuclear
   c. Solar thermal and hydro
   d. Nuclear and hydro

3. To efficiently carry voltage across long distances, power lines _____.
   a. maintain the same voltage from the generation station to the distribution substation
   b. carry steady high-voltage DC electricity
   c. use single-phase wiring
   d. use three-phase wiring

4. True or False: Some buildings may have more than one meter.
   a. True
   b. False

5. True or False: Energy storage capacity will need to increase to store the solar energy added to the power grid.
   a. True
   b. False

Module 57101 Review Questions

1. The *NEC*® requirements for PV installations can be found in _____.
   a. *NEC Article 422*
   b. *NEC Article 450*
   c. *NEC Article 517*
   d. *NEC Article 690*

2. In 2020, what was the average annual electricity consumption for a US residential utility customer?
   a. 893kWh
   b. 3,600W
   c. 10,715kWh
   d. 2MW

3. Which type of system is best to use in areas where access to the power grid is inconvenient or unavailable?
   a. Grid-access system
   b. Grid-interactive system
   c. Grid-tied system
   d. Off-grid system

4. When excess energy produced by a private solar PV system is credited to the power grid, it is known as _____.
   a. buyback metering
   b. grid metering
   c. net metering
   d. carryover metering

5. What is the main benefit of a grid-interactive system?
   a. The system's batteries provide energy when the solar array doesn't generate electricity.
   b. The system's energy solution may be the DC battery of an electric vehicle.
   c. The system is not dependent on the power grid.
   d. The system typically has five days of autonomy.

6. Which is a true observation about the Sun's path in Earth's sky?
   a. The Sun rises in the west and sets in the east.
   b. In the Northern Hemisphere, the Sun's path appears lower in the sky in the summer months.
   c. In the Northern Hemisphere, the Sun favors a southern trajectory.
   d. The solar window is bounded by the Vernal (spring) and Autumnal (fall) Equinoxes.

7. With true north as the reference point, in the Northern Hemisphere true south is true north plus _____.
   a. 0 degrees
   b. 90 degrees
   c. 180 degrees
   d. 270 degrees

8. Suppose a customer lives in an area with four peak sun hours per day, uses an average of 36kWh per day during the critical design month, and desires 50% replacement of grid power. What is the required system size?
   a. 1,500Wh/day
   b. 2,500Wh/day
   c. 3,500Wh/day
   d. 4,500Wh/day

9. When a motorized system is used to adjust an array to follow the motion of the Sun, it is known as _____.
   a. trailing
   b. trolling
   c. tracing
   d. tracking

10. Which method could you use to establish true south in order to know which roofing surfaces to plan for the solar array installation?
    a. Use a compass and magnetic declination to find true north and then add 90 degrees to find true south.
    b. Use a compass and magnetic declination to find true north and then add 180 degrees to find true south.
    c. Use a roof angle finder and optimize for the location's latitude.
    d. Use a sun path chart to find the solar window.

11. Which time of year is best for shade analysis?
    a. Spring
    b. Summer
    c. Fall
    d. Winter

12. Watt's law uses which three variables?
    a. Resistance, voltage, and current
    b. Power, voltage, and current
    c. Power, ohms, and current
    d. Ampacity, voltage, and current

13. What is the current when the voltage is 120V and the resistance is 10 ohms?
    a. 10 amps
    b. 12 amps
    c. 120 amps
    d. 1,200 amps

14. Wiring two 30V/5A PV modules in series produces _____.
    a. 30V/5A
    b. 30V/10A
    c. 60V/5A
    d. 60V/10A

15. When comparing motion systems, which has the highest energy yield?
    a. Dual-axis tracking
    b. Fixed tilt
    c. LIDAR tracking
    d. Single-axis tracking

16. True or False: Some PV modules can absorb reflected light on the back of the module.
    a. True
    b. False

17. PV module Standard Test Conditions (STC) testing provides a fixed set of indoor, laboratory conditions for rating PV modules and includes which of the following conditions?
    a. Solar cell's operating temperature of 35°F
    b. Solar cell's operating temperature of 35°C
    c. Incident solar irradiance level of $1000W/m^2$
    d. Air mass of less than 1.0 spectral distribution

18. Why is the PV module nameplate important?
    a. It provides the required battery amp-hours for the system.
    b. It indicates the maximum power output (STC).
    c. It indicates the inverter efficiency rating.
    d. It provides a lookup table to factor in temperature deviations from STC.

19. Which of the following safety features are commonly integrated into a string inverter?
    a. Lightning arrestor
    b. Ground-fault protection
    c. Remote disconnect
    d. Aluminum wiring

20. Which of the following is built into a microinverter?
    a. Lightning arrestor
    b. Ground-fault protection
    c. Rapid shutdown
    d. Aluminum wiring

21. Which of the following stores energy from the PV array to supply electrical loads as needed?
    a. Array disconnects
    b. Batteries
    c. Charge controllers
    d. Grounding systems

22. The best charge controller to use with sealed batteries is a(n) _____.
    a. MPPT charge controller
    b. PWM charge controller
    c. relay transistor charge controller
    d. shunt charge controller

23. Which of the following is a mechanical BOS component?
    a. Racking hardware
    b. PV module
    c. Inverter
    d. Combiner box

24. Which of the following is an electrical system BOS component?
    a. Charge controller
    b. Conduit
    c. Inverter
    d. Microinverter

25. Which of the following is a true statement about wires?
    a. Wires are capacitors wrapped in black insulation.
    b. Wires are rated in amp-hours.
    c. Wires are never stranded.
    d. Wire sizes are expressed in gauge numbers: the higher the number, the smaller the wire.

26. How many grounding locations are included in a solar design?
    a. One on the AC side
    b. One on the DC side
    c. Two on the DC side
    d. Two: one on the DC side and one on the AC side

27. What is the main purpose of a combiner box?
    a. To step down the voltage in the system
    b. To join wiring from multiple arrays into a single larger wire
    c. To discharge excess electricity from the system
    d. To conduct electricity

28. Where is the AC disconnect located?
    a. Below the D-ring on a fall protection harness
    b. In the basement
    c. At the microinverter
    d. Near the service panel

29. What do you need to take into consideration when assessing the required free length of a PFAS?
    a. ANSI safety rating
    b. Deployment of the shock absorber
    c. The system's energized state
    d. The terminal end sizing

30. Always check that you are using the right tool for the job, but in general, PV systems require the use of _____.
    a. Category I multimeters
    b. Category II multimeters
    c. Category II or III multimeters
    d. Category III or IV multimeters

31. Where in the PV system might an arc flash most likely occur?
    a. The AC side
    b. The DC side
    c. The AC and DC side
    d. It doesn't occur in the PV system.

32. Which type of battery releases hydrogen gas when overcharged?
    a. Flooded lead-acid (FLA)
    b. Gel
    c. Lithium-ion
    d. Sealed lead-acid (SLA)

33. Why is it important to educate the customer about energy efficient measures?
    a. To sell more services
    b. To decrease a building's energy consumption before sizing a solar system
    c. To decrease a building's energy consumption because it is environmentally the right thing to do
    d. All of these choices are correct.

34. To convert from kWh to Wh, multiply by what value?
    a. 10
    b. 100
    c. 745.7
    d. 1,000

35. Which of the following is *not* a task performed during a site assessment?
    a. Inspection of the interior and exterior of a building
    b. Site research
    c. Sketching a preliminary site survey of the jobsite's buildings
    d. System commissioning

36. What document does solar PV system design provide for product purchase and inventory?
    a. Bill of Materials
    b. Electrical wiring diagram
    c. List of Materials
    d. Product list

37. What professional qualifications are required for a worker to set up fall protection, scaffolding, and barriers?
    a. ANSI certification
    b. NABCEP Associate
    c. OSHA-30
    d. Completion of a certified competency in that area of expertise

38. True or False: Ballast mounts use concrete blocks to weigh down a racking system.
    a. True
    b. False

39. Which of the following is often the result of wiring too many PV modules in the same circuit?
    a. Heat fade
    b. Bypass diode failure
    c. Inverter losses
    d. Burnt terminals

40. Which of the following is a point of interconnection between a commercial or residential building and the power grid?
    a. Electrical meter or smart meter
    b. Microinverter
    c. Service panel
    d. Transformer

41. Three-phase wiring carries _____.
    a. AC power
    b. DC power
    c. low-voltage power
    d. 80° phased current

42. Which of the following is *not* a power-grid distribution component?
    a. Main electrical service panel
    b. Power lines
    c. Power poles
    d. Step-down transformers

43. Why will battery storage capacity need to increase?
    a. To store the solar energy for distribution at any time of day
    b. To regulate the distribution of solar energy within the power grid
    c. To transform solar energy to electrical energy
    d. To reduce the amount of time it takes for electricity to be distributed from the utility company to the customer

CORNERSTONE OF CRAFTSMANSHIP

Tony Johnson
Power52 Foundation
Craft Instructor

How did you choose a career in the industry?

When I left the military, my background included maintenance activities on cargo aircraft including the jet engines. I used that knowledge to apply for a position as a turbine/generator mechanic in the generation side of an electrical utility company. Throughout my career, we were always looking for different ways to "green" the process of electrical generation. Solar power generation has been making technical advancements that need to be taught.

Who inspired you to enter the industry?

My grandfather inspired me to enter the industry. He was a Master Electrician with the same company that I retired from. From the time that I could hold tools, he taught me how to properly use them and the theory behind why they are used.

What types of training have you been through?

I have NCCER certifications in Industrial Maintenance Mechanic, Millwright, Power Generation Maintenance Mechanic, Mobile Crane Operations, Advanced Rigging, Your Role in the Green Environment, Solar Photovoltaic System Installer, and all of the instructor certifications for the aforementioned modules. I also have NABCEP Photovoltaic Installer Instructor Certification, OSHA 500, and managerial training.

How important are education and training in construction?

Education and training are very important, not only the initial training but continual education as well. Construction training is extremely important in the Solar Industry. The technology is rapidly changing, and the electrical codes are being updated to keep pace with the technical changes.

How important are NCCER credentials to your career?

NCCER credentials have given me multiple advancement opportunities and career path changes to follow. NCCER credentials are accepted internationally and by NABCEP as formal education. They are well recognized in the construction and renewable energy industry.

How has training/construction impacted your life?

The construction industry and the training in this field allowed me to lead a productive life and raise a family in a career that I love.

What kinds of work have you done in your career?

I have been a mechanic on turbine generators in the fossil, nuclear, biomass, wind, and hydroelectrical generation. I became a supervisor managing those mechanics. After that I became a craft instructor for those fields as well as an equipment operations instructor.

Tell us about your present job.

I am a craft instructor for Power52 Foundation. We are delivering NCCER's Core, Your Role in the Green Environment, and Solar Photovoltaic Systems Installer curricula. The demographics we serve are graduating high school seniors with no plan, historically marginalized and at-risk people, as well as displaced workers. We not only supply them with the technical skills needed for a successful career, but also life skills. We have also partnered with several higher educational institutes to assist our graduates with continual education opportunities.

What do you enjoy most about your job?

The thing I enjoy most about my job is helping someone change paths and supplying the education needed for a successful career. I also enjoy watching the trainees gain the knowledge and confidence to take them to the next step. It is very fulfilling when graduates return and share their testimony of the journey that they have taken and how they are living the dream.

What factors have contributed most to your success?

Embracing change, never becoming complacent or accepting the *status quo*, and always striving to become better are all factors that have contributed to my success.

Would you suggest construction as a career to others? Why?

I would absolutely suggest construction as a career to others because there will always be a need for the construction field. Not only is the construction field lucrative, but it also offers advancement opportunities with no need for a college education. In construction, you start out with a blank canvas and when the project is complete, you will always be a part of that project. You can show people and say with pride that you helped build that.

What advice would you give to those new to the field?

If you are new to the field, you should never, ever stop learning new things. Keeping your skill set relevant to current industry needs is critical. A wide range of paths will present themselves as you advance in your career and life. Never be apprehensive to try something new, and if you fail, try it a different way until you succeed. Lastly, follow your dreams.

Interesting career-related fact or accomplishment:

I never imagined that I could have achieved what I have thus far in my life and career without having a college education. You never know until you try.

How do you define craftsmanship?

I believe craftsmanship is a feeling or act of an individual performing a task or set of tasks that, when complete, brings a thought process to an observer that whoever did this project 100 percent loves what they do.

Answers to Section Review Questions

Answer	Section	Objective
Section One		
1. d	1.1.0	1a
2. d	1.1.1	1a
3. b	1.2.0; *Table 1*	1b
4. d	1.3.4; *Table 3*	1c
5. c	1.4.2	1d
6. a	1.4.3	1d
7. d	1.5.0	1e
Section Two		
1. b	2.1.2	2a
2. b	2.1.3	2a
3. c	2.2.0; *Table 6*	2b
4. b	2.3.1	2c
5. c	2.4.3	2d
6. a	2.5.1; *Table 8*	2e
Section Three		
1. c	3.1.0	3a
2. a	3.2.1	3b
3. d	3.2.4	3b
Section Four		
1. b	4.1.0	4a
2. c	4.2.0	4b
3. b	4.2.3	4b
Section Five		
1. d	5.1.0	5a
2. d	5.2.0	5b
3. c	5.3.6	5c
4. c	5.4.3	5d
5. a	5.4.5	5d
6. a	5.5.2; *Table 19*	5e
Section Six		
1. c	6.1.0	6a
2. b	6.1.1	6a
3. d	6.1.2	6a
4. a	6.1.3	6a
5. a	6.1.4	6a

Answers to Odd-Numbered Review Questions

Answer	Section	Objective
1. d	1.1.1	1a
3. d	1.3.1	1c
5. a	1.3.3	1c
7. c	1.4.2	1d
9. d	1.4.3	1d
11. d	1.4.4	1d
13. b	1.5.2	1e
15. a	2.1.1	2a
17. c	2.3.1	2c
19. b	2.4.0	2d
21. b	2.5.1	2e
23. a	3.1.0	3a
25. d	3.2.1	3b
27. b	3.2.4	3b
29. b	4.1.0	4a
31. b	4.2.2	4b
33. d	5.1.0	5a
35. d	5.2.0	5b
37. d	5.4.0	5d
39. d	5.5.2; *Table 19*	5e
41. a	6.1.2	6a
43. a	6.1.4	6a

GLOSSARY

Air mass: The volume of air that sunlight passes through to reach Earth.

Alternating current (AC): An electrical current in which the direction of electron flow switches back and forth in cycles.

Altitude: In relation to solar PV, the Sun's elevation in Earth's sky in relation to Earth's horizon, measured in degrees.

Ambient temperature: The air temperature of an environment.

Amorphous: A low-efficiency type of photovoltaic cell characterized by its ability to be used in flexible forms, like thin film.

Ampacity: The maximum current in amperes a conductor can carry continuously under the conditions of use without exceeding its temperature rating.

Array: Multiple solar photovoltaic (PV) modules wired together to make a larger structure.

Autonomy: The number of days a fully charged battery system can supply power to a building's critical loads without recharging.

Azimuth angle: In relation to solar PV position, the angle clockwise from true north for setting the left-right rotation of a PV module.

Back sheet: The back of a solar photovoltaic (PV) module.

Backfeed: The flow of electric power in the direction reverse to that of the generally understood or typical flow of power.

Balance of system (BOS): The additional parts which connect the primary components together and are required to execute a solar PV installation based on the engineer's solar plan.

Brownout: A temporary decrease in grid output voltage typically caused by peak load demands.

Building-integrated photovoltaics (BIPV): A PV system built into the structure as a replacement for a building component such as roofing.

Bypass diodes: An electronic part wired in parallel with individual PV modules to direct current around a poor-performing PV module so overall array performance is less affected.

Charge controller: A device used to regulate the charging and discharging of a battery system to prevent overcharge and excess discharge.

Clean energy: An energy source that doesn't emit greenhouse gases.

Combiner box: An electrical distribution box with DC safety overprotection used to combine multiple sets of strings of PV modules into one set of conductors via a fused terminal block.

Declination: The angle between Earth's Equator and a celestial object, like the Sun.

Depth of discharge (DOD): The percentage of a battery's capacity that may be safely used to power a load without damaging the battery.

Direct current (DC): An electrical current that flows in one direction.

Distributed energy resources (DER): Independent, renewable energy source generators, such as residential solar PV systems.

Doped: Having specific impurities added to a material to produce a positive or negative charge.

Elevation: The height of something in reference to a given point.

Energy efficiency: Reducing a building's energy consumption by using appliances and lighting solutions that are designed to efficiently use energy.

Grid-interactive system: A grid-tied system with an energy storage solution.

Grid-tied system: A solar PV system that operates in parallel with the power grid, providing supplemental power to the building but unable to be autonomous from the system.

Heat fade: A condition in which a PV system operates inefficiently during periods of high heat.

Home run: Main collection of wires that connect a fixture to a centralized location for power distribution.

Hybrid power system: A solar PV system with multiple renewable energy sources, such as hydro power, solar photovoltaics, solar thermal, wind turbines, or generators.

Interconnection: The technical and procedural matters associated with operating interactive PV systems and other distributed generation sources in parallel with the electric utility system, as defined by NABCEP.

Inverter: A device used to convert direct current to alternating current.

Latitude: A measurement expressed in degrees for a location's distance in relation to the Equator.

Magnetic declination: The angle in degrees between true north and magnetic north that is specific for an observer's location on Earth.

Maximum poswer point tracking (MPPT): A battery charge controller that provides precise charge/discharge control over a wide range of temperatures.

Monocrystalline: A type of PV cell formed using thin slices of a single crystal and characterized by its high efficiency.

Net metering: A utility program enabling customers to credit their solar energy surplus into the power grid to draw upon later.

Net-zero building: A building that generates and consumes the same amount of energy within a year, resulting in a net use of zero.

North American Board of Certified Energy Practitioners (NABCEP): A volunteer board of renewable energy system professionals that provides standardized testing and certification for PV system installers.

Off-grid system: A solar PV system with an energy storage solution that is independent of the power grid. Also known as a *stand-alone system*.

Peak sun hours: Solar irradiance a given location receives when the Sun's intensity is highest for a specific number of hours.

Polycrystalline: A type of PV cell formed by pouring liquid silicon into blocks and then slicing it into wafers. This creates non-uniform crystals with a flaked appearance that have a lower efficiency than monocrystalline cells.

Pulse-width modulation (PWM): A method of turning off and on the power in a circuit to decrease current.

Renewable energy: An energy source that quickly replenishes the supply of energy.

Semiconductor: A material that exhibits the properties of both a conductor and an insulator.

Solar irradiance: The amount of solar radiation that strikes Earth's surface in a given span of time (kW/m^2/days).

Solar photovoltaic (PV) cell: A semiconductor device that converts the energy of sunlight into direct current. Also known as a *solar cell*.

Solar photovoltaic (PV) module: A PV system component consisting of numerous electrically and mechanically connected PV cells encased in a protective glass or laminate frame. Also known as a *PV panel*.

Spectral distribution: The distortion of light through Earth's atmosphere.

Standard Test Conditions (STC): Controlled laboratory environment specifications for PV module operating temperature, solar irradiance, and air mass used in standardizing PV module performance ratings.

Sun path: The observed movement of the Sun across Earth's sky.

Thin film: A low-efficiency type of photovoltaic cell characterized by its ability to be used in flexible forms. Also called *amorphous*.

Tilt angle: In relation to solar PV, the up-down tilt position of a PV module or array.

Utility-scale solar generating systems: Large solar farms designed to produce power in quantities large enough to operate a small city.

Voltage: The electric pressure between two points, originating at an electrical circuit's power source, which pushes electrons through a circuit.

Watt-hours (Wh): A unit of energy that describes the amount of power used over time, typically used for metering.

ADDITIONAL RESOURCES

This module presents thorough resources for task training. The following reference material is suggested for further study.

Articles/Policy

- "Best Practices for Operation and Maintenance of Photovoltaic and Energy Storage Systems; 3rd Edition," National Laboratory of Renewable Energy: *https://www.nrel.gov/docs/fy19osti/73822.pdf*
- Interconnection: IEEE 1547 and 2030 Standards for Distributed Energy Resources Interconnection and Interoperability with the Electricity Grid (nrel.gov)
- UN Paris Agreement: *https://www.un.org/en/climatechange/paris-agreement*
- US climate change policy: *https://www.whitehouse.gov/wp-content/uploads/2021/10/US-Long-Term-Strategy.pdf*

Tools

- "Appliance Energy Calculator," US Department of Energy: *https://www.energy.gov/energysaver/maps/appliance-energy-calculator*
- Magnetic declination lookup, National Oceanic and Atmospheric Administration (NOAA): *https://www.ngdc.noaa.gov/geomag/calculators/magcalc.shtml*
- NREL SolarAPP+ Inspection checklist: *https://solarapp.nrel.gov/docs/approval-document-pv-example.pdf*
- Solar irradiance lookup, National Renewable Energy Laboratory's (NREL): *https://www.nrel.gov/gis/solar-resource-maps.html*
- Sun path chart lookup, University of Oregon Solar Radiation Monitoring Laboratory: *http://solardat.uoregon.edu/SunChartProgram.html*
- US Department of Energy: How Does Solar Work? *https://www.energy.gov/eere/solar/how-does-solar-work*